Personal and Planetary Wellness

Addressing Climate Change, Health, and Social Justice Challenges

Marilyn Cornelius

ISBN: 1547278080
ISBN-13: 978-1547278084

DEDICATION

This book is dedicated to Sidharth, my unborn nephew.
It is also for every child in the world, of every species. We will create a secure future for you. We must.

Table of Contents

ACKNOWLEDGMENTS

Diane and Margaret were my intrepid editors, advisors, and proofreaders for this book – any errors are mine alone. There is no way this book would have been assembled so expeditiously without their dedication and persistence. Thank you to you both for your invaluable suggestions and feedback, your supportive energy, and undying love for me and for our shared devotion to a safer and happier world. Margaret, who is my mom and (an)altruist at Alchemus Prime, wrote some of the wellness essays in her beautiful and nurturing voice. I am grateful to the scientists and researchers whose shoulders we stand on as we explore their findings and aim to communicate them in simpler terms. It is a privilege to pursue my questions and cultivate ways for living well personally and for the planet. I am thankful to Moshe, who told me, a few years ago, to write and share my thoughts - I have not stopped since! Most of all, I am thankful to Ma Nature for her inspiration, which drives all my work, turning it into play.

NOTES:

I created all the diagrams and took all the pictures in this book, except for my photo on the back cover, which Ike Shin captured. The front cover photo represents the Alchemus Prime Diamond Model and one of our slogans: **connect, create, transform, sustain.** We connect with ourselves and nature (through meditation and time spent outside); create and implement new habits, such as what we eat (using design thinking); transform our behavior and our very selves (applying behavioral sciences), like a caterpillar turns into a butterfly; and sustain those earth-friendly changes over time (using lessons from nature via the discipline of biomimicry).

Introduction

I am a seeker. The status quo never satisfies me. I want a world that is safer, healthier, and more equitable. Don't you? In my search for answers, I have often found amazing insights and often, profound questions. This book contains a collection of 70 essays out of over 250 I've written in the last two and a half years that exemplify my seeking. The essays focus on how to engage with our loved ones, our colleagues, nature, and ourselves in ways that create harmony. Each essay begins with either a piece of research or a direct experience, and aims to distill insights and principles for how we might become better leaders in a constantly changing, sometimes depressing, and sometimes scary world. These essays started off as blogs and have been reformatted and updated for this book.

For me, the elephants in the room with regard to our current society are climate change and lifestyle diseases. I look around and see people who are suffering, a global society that is precarious in its politics, and animals and plants under constant assault from our activities. It turns out that climate change and our illnesses are intricately related through our daily actions: what we eat, how we transport ourselves, and many other decisions we make as active and sometimes passive consumers. In a time of disenchanting politics, it's more urgent than ever to focus on what each of us can do to make a direct positive impact on human society, a safer future for our children and theirs, our forests, oceans, grasslands, other ecosystems, the climate, and public health.

Social justice is a third elephant in the global room. It is a thread that ties together what we are seeing in terms of ill health and environmental damage. All too often, negative health and environmental impacts happen to the poor, minorities, and disadvantaged people, and to animals that cannot defend themselves. Just as Gandhi resisted the British by empowering locals to make their own clothes, people everywhere can resist the oppression of animals, the poor, children, women, people of color, LGBTQ communities, and other oppressed groups by making small, targeted changes in their daily routines. In this book, I explore how to address these issues in practical ways.

This book gives you tips on how to:
- As an individual, contribute to your own wellness and that of the Earth, starting with one action at a time;

- Be a leader and role model in your own home or workplace for wellness, efficiency, and creativity;

- Become more mindful and aware of yourself, people around you, and nature; and

- Be more confident and better aligned with your values as an authentic leader.

The evidence covered in this book spans the last few years and includes:

- What each and every individual can do about climate change, by taking charge of daily actions such as diet and exercise.

- How to improve your health using three tactics: what you put into your body, how you move your body, and how you relate to other living beings.

- How you can connect with nature in ways that improve health, spark boundless creativity, and support increased productivity.

- How to be a leader in your current life, professionally and personally through mindfulness and heightened awareness and active overcoming of social injustice.

- How diets that contain high amounts of meat and dairy worsen climate instability, water pollution and scarcity, deforestation, biodiversity loss, food insecurity, animal cruelty, poverty, and in turn, our own health, leading to lifestyle diseases such as obesity, diabetes, cancer, and heart disease.

Each chapter offers a set of perspectives on a particular aspect of personal and planetary wellness. Chapter 1 provides an overview of the framework I have been developing, which I call Personal and Planetary Wellness. It contains the ways in which our daily choices affect our own bodies and longevity, as well as global resources such as water, food, forests, wildlife, and more.

In Chapter 2, Margaret, a medical doctor and (an)altruist at Alchemus Prime, and I explore the evidence for how the foods we choose to eat can either make us very ill, or prevent or reverse lifestyle diseases such as cancer, diabetes, obesity, and heart disease, without drugs. Chapter 3 explores the myriad connections between what we eat, and how it contributes to or reduces climate change. Some of the medical education themes Margaret addresses in Chapter 2 are broadened in Chapter 4 into a discussion of how we might improve the entire education system to better nurture people and protect the Earth.

In Chapter 5, I discuss racism and social justice based on my personal experiences as a woman of Indian descent who was born and raised in Fiji, spending half of her life in California falling in love with the cultural diversity in the San Francisco Bay Area, and working on climate change and wellness issues. I also explore social justice in prison, from a genetic perspective, and against animals.

Chapter 6 contains explorations of the benefits meditation provides, which extend far beyond stress-reduction. There are benefits to productivity, longevity, brain function, and much more. I discuss different meditation techniques and how to apply them in the workplace and in our personal time.

Chapters 7 and 8 explore innovation. In Chapter 7, I delve into biomimicry, a discipline that focuses on learning and applying the wisdom of nature's processes, designs, and solutions. Biomimicry can teach us so much about how we design systems to engage people at work, how to become more mindful, how we integrate solutions for better results, and so much more. Chapter 8 contains essays about design thinking, a process for fostering systematic creativity in problem solving. I examine design thinking and its value, how to scale it, and how to integrate it with other methods, including behavioral science techniques, to achieve more effective results for personal and professional challenges.

Chapter 9 focuses on behavior change, which is my specialty and passion. I walk through different steps we can take to be better motivated, stop procrastinating, and create more alignment between our values, thoughts, goals, and actions. In Chapter 10, I apply behavioral science principles to climate change and sustainability issues, highlighting how changing our own actions can bring us quick and effective solutions as we fight climate change.

I conclude with Chapter 11, which investigates integrity in leadership. This chapter contains essays that dig deeply into the true self, and how to lead in ways that maintain the core of who we are, by aligning values, thoughts, and actions, as we aim to do at Alchemus Prime. In my professional and personal experience, I have found that nothing of value can be created or sustained without integrity, so I end with the most important value I know.

In sum, these short essays offer guidance in what I hope is accessible language, each with a reflection at the end that prompts you to think a little more deeply about how you can be a stronger force for change. You could read this book from start to finish, or pick any essay at will and see where it takes you – see the Table of Contents for essay titles. Check out the Glossary for any terms that are unclear to you. Use the Chapter Notes if you'd like to learn more about any of the topics I mention – they contain links to peer-reviewed research papers, various types of scientific and policy reports, videos, and news articles.

Most importantly, this book is not just for reading; it's a call to action. If you really care about climate change, lifestyle diseases, and social justice, and how they are affecting you and your family, but aren't sure what exactly you can do about it, this book is for you.

Lastly, I can't promise you that reading this book will be comfortable. You may encounter feelings of defensiveness, incredulity, doubt, skepticism, and more, and I urge you to persist in getting answers to your questions, asking more questions, and understanding why your actions matter so much. Once you discover the power of your daily choices, you may want to explore how you might differently choose the consequences you want to support. One choice in particular makes the biggest difference: what we eat. Read this book to find out why.

Chapter 1: Personal and Planetary Wellness - The Big Picture

Personal and Planetary Wellness: Winning

When I was in Arizona interviewing teachers from Navajo Nation about the actions they see as most important to enhance indigenous learner success, one of the stories I heard really stayed with me. It was about a particular melon species that became submerged when a dam was built on lands that belonged to a particular American Indian tribe. The dam, and the subsequent loss of melons, changed the course of that tribe's language and culture; the melon could no longer be used for food or ceremony, and eventually the word for it was lost for lack of use.

This story gave me pause: here was an example of a direct relationship between nature's integrity and human wellbeing. A change to *one* species in an ecosystem had directly and negatively affected some aspects of human culture and language practices, leading to the extinction of both!

Another example came to light during the course of my research[1] and activism[2] on climate change; I work to make visible the harmful connections between animal agriculture and climate change. The raising, processing, refrigeration, transportation, storage, and consumption of animals for the production of beef, chicken, pork, cheese, and other animal products create more greenhouse gas emissions[3] than any other industry, including all of transportation. Climate change harms our most vulnerable. our children. Children's asthma worsens[4] from the air pollution caused by cars. Many species face extinction[5] from rapidly increasing temperatures, reducing the planet's resilience.

Aside from adverse climate impacts, there are negative health outcomes from eating animal products too. A *Journal of American Medical Association* (JAMA) article[6] found that eating red meat increases the risk factor for diabetes by a staggering 48%.

Put simply, what is bad for humans is bad for the planet, and the reverse is true too.

Fortunately, we can change this relationship for the better so everyone wins. A plant-based diet is the most important action we can take to address climate change, and to reverse diabetes[7] and heart disease.[8] Veganism can also make a difference when it comes to saving rainforests[9] and addressing animal cruelty[10] in factory farms.

Another win-win for climate change lies in changing our other daily behaviors. In addition to what we eat, how we move, and how we stay warm and cool really matter for the climate. If we walk or ride our bicycles instead of driving, we avoid greenhouse gases altogether, and improve our health through the exercise. If we use evaporative cooling by wearing a wet T-shirt instead of using air-conditioning on a hot day, we use less energy and emit fewer greenhouse gases. If we wear warm clothing instead of cranking up our heaters, we become more acclimated and use less fossil-fuel-based energy. Although climate change strategies are typically divided into mitigation (reducing climate change) and adaptation (coping with the effects), recent research[11] suggests that adaptive behaviors can be very effective options for mitigation as well.

Indigenous peoples, such as the Navajo teachers I visited, know how to live in harmonious and resilient ways that adapt to changing environmental conditions. It is essential to preserve indigenous knowledge and practices so that humanity can restore and regenerate the earth. However, each one of us, no matter where we come from, can make a difference every day when we make decisions about what to eat, how to get from point A to point B, and how to get a little warmer or cooler using clothing. Turning the climate crisis around by changing our everyday behaviors is a win-win for human and planetary wellness. Why not?

Reflection:

What are some examples from your life that illustrate the connection between taking care of yourself and taking care of the planet?

Example from our friend Diane: Gardening: Keeps me healthy with fresh produce and avoids supporting industrialized agriculture which pollutes the planet.

1.

2.

3.

The Rise of Common Sense

"The new common sense is about knowing that all of life is significant and that you must take care of it."

-- Barbara Marciniak, *Path of Empowerment*

As I celebrated the New Year in Fiji with my parents, I read in the news that a prominent businessman, 52, had collapsed and died from a massive heart attack. I was reminded that our inner climate, if you will, is in crisis: we suffer from heart disease, obesity, diabetes, and cancer in epidemic proportions. Diabetes is actually the number one killer in Fiji, this tropical paradise where I was born.

Our bodies are experiencing chaos. The world's climate mirrors this crisis state.

The global average temperature of the earth has been rising; 2014 was slated to be the hottest year[12] on record, according to NASA. Sadly, 2015 and 2016 broke that record.[13]

The lungs of the earth, our forests, are being destroyed to graze cattle and plant genetically modified corn and soy, which we eat in the form of burgers, sugary snacks, and sodas containing high-fructose corn syrup. It's no wonder we are ill. We continue to pollute our air, water, and land with chemicals, toxins, pesticides, hormones and so on, and then we wonder why we suffer from cancer,[14] or why our children are born with defects.[15] We torture and slaughter drugged animals in factory farms and ingest their trauma, their antibiotics,[16] and hormones. Consequently violence, depression and other forms of physical and mental illness are prevalent in our societies today.

As you ponder New Year's resolutions each year, ask yourself: what will I do this year to reduce the (inner and outer) chaos?

Fortunately, I believe this negative feedback loop is about to surrender to common sense: because we are connected to all living systems, any harm we do to the earth, we do to ourselves. More and more human beings are becoming aware of the direct connections between what I call personal and planetary wellness.[17] We are beginning to act on what we know. Evidence for the rise of common sense is readily available through scientific research, popular culture, animal rights activism, startups, as well as wellness movements.

9

Look up Dr. Gabriel Cousens, and see what reverses diabetes[18] (hint: it's not drugs). Pick up The China Study[19] or The World Peace Diet[20] and read about the myriad connections between our diet, lifestyle disease, and planetary wellness. Watch documentaries such as Cowspiracy,[21] Forks over Knives,[22] and What the Health[23] to get a sense of what is at stake for the climate and our own bodies if we don't change what we eat. If you're really brave, watch Earthlings[24] (it's not for the faint of heart or weak-stomached).

The facts are all around us. If you peruse the evidence with an open mind, the solution becomes crystal clear: a shift away from meat and dairy is best for all life to thrive on Earth. If we want to heal our climate, all living beings, and ourselves, we need to pay attention to what we put into our bodies, where it comes from, how it reaches us, and how it's processed.

If you are motivated to change your diet, you might want look up Physicians Committee for Responsible Medicine's (PCRM) 21-day Vegan Kickstart[25] as an excellent way to start any New Year. Veganuary[26] is another program that is taking off; it focuses on becoming vegan to reduce animal suffering. Or, you can do what my father does: eat meat only once a day, and buy most of your food from a produce or farmers market.

Another pathway to becoming more aware and healthy is through meditation and yoga. I highly recommend Transcendental Meditation[27] (TM), Mindfulness-Based Stress Reduction[28] (MBSR), and a yoga practice; and perhaps you're ready for a Vipassana[29] retreat. Meditative practices cleanse your mind, allowing your inner knowing to emerge. For instance, is it safe or fair or healthy to repeatedly impregnate cows and when they give birth, keeping calves just out of reach of their moms, so we can use their milk? Do we pause to consider this each time we eat pizza or ice-cream? What other unexamined choices are we making daily?

If there has ever been a time for facing the decisions we may not directly make, but are directly supporting with our consumer choices, that time is now.

If you're into tracking your consumer choices, you can do so in a fun way with Oroeco's platform. Oroeco[30] lets you track your everyday choices so you can see how they align with your values, and reduce climate change.

If your way of wanting to do your part is by raising your voice for the animals, you may want to join your local chapter of an animal rights group[31] such as Mercy For Animals, or vegan group, such as Vegan Outreach. To alert climate leaders such as Al Gore about the harmful connections between animal agriculture and climate change, join Operation Missing Link. The existence of all these relatively new organizations is another indication that common sense is taking root.

If you are making bold changes this year, and would like to meet people who share your values and are taking similar actions, join your local vegan, animal rights, justice, or other groups, as well as Facebook communities. You'll find that you are not alone. This rising tide of human consciousness is global, and luckily for all of us, it's accelerating.

We humans, with our infinite capacity to hear the call of common sense, are the solution. It doesn't matter if you're into incremental, one-step-at-a-time solutions (Meatless Monday,[32] anyone?) or holistic, systemic change (going vegan today), every positive action counts. The best time to take action is now. What will you do today?

Reflection:

What are some other common sense solutions you can think of to fight climate change?

Example from Margaret: I avoid high-energy appliances like the dryer by hand washing my clothes and drying them in the sun.

1.

2.

3.

Going to Hell in a Hand Basket* or Converging on a Solution?

Inspired by an article[33] on indigenous ethics, I was reflecting on the sacred connection indigenous peoples share with plants, animals, stars, and planets.

What about the rest of us? Trapped in crowded, tech-saturated urban areas with busy-bee lives, do we know balance? Do we know reverence?

Stunningly, the root of many of the world's environmental and health ills are embedded in the way we have developed systems of animal agriculture.

Forests are being removed in the Amazon and other regions of the world to graze cattle. Biodiversity is soon lost as these forest habitats are destroyed. The land becomes degraded, and rainfall, with no remaining opportunity for filtration through the roots of the forest, washes away precious topsoil, polluting rivers and other waterways.

As a consequence of animal agriculture, food security for the indigenous and the poor are threatened, as only those who can afford meat and dairy purchase it. Cattle and other animals such as chickens and pigs raised in factory farms are fed grain that could otherwise be used to sustain starving humans around the world. Those of us on the other end of the spectrum, who are not starving, are eating ourselves to death. Cancer, type 1 and 2 diabetes, heart disease, and obesity are all linked to high levels of meat and dairy in the typical Western diet.

As if this wasn't egregious enough, life cycle analysis of animal agriculture (that includes grazing, refrigeration, processing, transportation, storage, etc.) indicates that it is the largest contributor to global greenhouse gas emissions. In fact, a recent report[34] states that if the world continues to adopt the typical Western diet, we could be looking at a 4-degree Celsius global temperature rise instead of a 2-degree rise – not a future I want to tell stories to my nieces and nephews about. Not a future they deserve…

In the previous essay, I highlighted some of the research and important sources to support these points. I want us to simply pause now and reflect on the system we have created or supported. Clearly, it isn't working. It is based on violence, hierarchy, hubris, greed, selfishness, shortsightedness, and ignorance.

Fortunately, this destructive paradigm is already crumbling. Reversing it, we arrive at veganism – refraining from eating or otherwise using any animals or animal products. A solution that allows forests to be restored, bringing back biodiversity, nurturing valuable soil, and filtering water into pristine aquifers as only nature can. Remember, forests are the most stabilizing force for our planet's climate.

A healthy vegan diet may reverse diabetes, heart disease, and obesity, and can prevent the triggering of cancer (if you don't think so, it really is time to read The China Study). As we show compassion to animals and release them from torture and cruelty, we practice less violence and more love to our fellow humans, dissolving the 'us and them' illusion so that all humans may eat lower on the food chain and live well. Ultimately, a vegan lifestyle honors life and allows living systems to thrive, placing humans in a more humble and compassionate position relative to all other forms of life.

If we take an in-breath, and become present to the life force that enters our body in the form of oxygen, then breathe out, releasing the carbon dioxide that can be toxic to us in large amounts but is beneficial to plants, we become aware of how connected we really are to all of life. And, how vulnerable we can be. Is there a loved one next to you, breathing in the air you just breathed out? As I sit here with my parents, I am all too aware of why I am fighting for a stable climate – for them, their grandchildren, and all the human and other animals who deserve a life of wellness and freedom.

Climate change, and all the interrelated challenges that stem from animal agriculture, are ripe for transformative human intervention. These questions remain with me:

- What tools and systems might we build that promote and sustain life?
- How might we treat all life with reverence, including our own?
- How might we begin now, with our own unique talents and skills, to serve all life?
- How might we collectively embrace veganism an optimal solution for healing Ma Nature?

What are your questions?

*An expression my late mentor, Steve Schneider, used often and only half-jokingly…

Reflection:

What are the most compelling reasons you would fight for a stable climate?

Example from Marilyn: To secure the future of my nieces and nephews and to help all children and animals be safe.

1.

2.

3.

Integration: Biomimicry, Veganism, and Justice

I am a big fan of the Bioneers National Conference. Last year, when I returned from this mecca of wonderful biomimics, I wrote prolifically about the highlights and questions I took away (See Chapter 5). I wrote another post for a company I was consulting for at the time. One of my readers, Diane, responded to that article with a set of important questions.

Essentially, the questions were about the intersections between biomimicry, veganism, and fair labor practices. Some of the specific questions from Diane, a vegan and animal rights activist, are listed below:

- Does a company that has biomimicry at heart avoid products that come from animals and/or are tested on animals?
- Is it important to favor recycled materials in biomimicry?
- Do biomimetic companies use labor that is equitable and fair to the workers as opposed to using people who have very low wages to maximize profits?

These questions struck me as central to the essence of the Alchemus Prime Diamond Model,[35] because they reach the core of our approach to ensure every life form wins through our work. Briefly, the Alchemus Prime Diamond Model integrates behavioral sciences (embracing change), design thinking (applying creativity), biomimicry (emulating nature), and meditation (being present and mindful) to address climate change and wellness challenges.

Our model leverages the deep intertwined relationship between climate change and wellness: human addiction to meat and dairy, which drives animal agriculture, is the leading contributor to climate change, when viewed from a life cycle analysis perspective.[36] The same behaviors we adopt for reducing climate change (i.e. reducing or eliminating meat and dairy from our diets) can alleviate the top lifestyle disease epidemics[37] facing humans around the world: diabetes, obesity and heart disease, and more.

The added benefit of eating a plant-based diet is the elimination of animal use, freeing up land to grow vegetables, grains, and roots to feed starving humans, and restoring some agricultural land to forest, allowing us to sequester water and start to reverse desertification and climate change.

Back to Diane's questions, which, when combined, basically ask: Shouldn't we practice biomimicry in ways that eliminate waste and are fair to humans and other animals? YES. Absolutely. Applying the Alchemus

Prime Diamond Model, we can assess the avoided or reduced greenhouse gas impacts, and harm to human health and labor practices, as well as measure the benefits to humans and nature from a particular project or intervention. Using this comprehensive analysis, we can tell whether any initiative is a true win-win, or simply creating benefit in one realm while generating harm in another.

Diane's questions point us to the holy grail of an integrated framework for measuring the benefit of any project in any domain: do the outcomes from one methodology create benefits and/or reduce harms in combination with all other methodologies and technologies being used, and overall for life? Biomimicry alone isn't enough, because although it aspires to create conditions conducive to life, nature demands killing and destruction of some species, individuals, and ecosystems for regeneration of new life forms. This is why we must integrate with other methods and include ethical parameters.

A truly powerful opportunity in the hands of humanity now is to leverage our ways of knowing, especially indigenous ways of knowing and practicing harmony with all life (what I call social biomimicry), and integrate them with other proven methods, such as behavioral sciences, to innovate our way out of the climate and wellness messes we've created. Our biggest asset is the human ability to change how we think and what we do.

Thank you Diane, for inspiring us; you make our work more robust.

Reflection:

Survey the products in your home. How many are not tested on animals, recyclable or biodegradable, and fair trade? List them here. Which meet all these criteria? Put a star next to those. What might you change in terms of your consumer choices?

Chapter 1 Notes

---

[1] This is one of my dissertation research projects I conducted with a team when I was at Stanford University, on how to reduce our greenhouse gas emissions by changing our behavior related to home energy use, transportation, and food. The peer-reviewed paper can be found here: http://link.springer.com/article/10.1007%2Fs12053-013-9219-5

[2] I am an activist with Operation Missing Link, which seeks to elucidate the harmful connections between animal agriculture and climate change: http://peoplepoweredpeace.org/operation-missing-link

[3] Robert Goodland and Jeff Anhang's paper, "Livestock and Climate Change" uses a life cycle approach to show that animal agriculture accounts for up to 51% of global greenhouse gas emissions. While this work was disputed, the authors successfully rebutted their skeptics. The paper can be accessed here: https://www.worldwatch.org/files/pdf/Livestock%20and%20Climate%20Change.pdf

[4] In California, pollutants like nitrogen from agriculture and transportation contribute to asthma, and climate change worsens the problem by converting the nitrogen intro nitrous oxide and ozone, which lead to cardiovascular and respiratory diseases: http://www.scientificamerican.com/article/climate-change-is-bad-news-for-california-children-with-asthma/

[5] Climate change increases the risk of extinctions: http://www.nature.com/nature/journal/v427/n6970/full/nature02121.html

[6] The article can be found here: http://archinte.jamanetwork.com/article.aspx?articleid=1697785

[7] The Physicians Committee for Responsible Medicine (PCRM), has summarized studies that show how meat and dairy inhibit insulin's functions: http://www.pcrm.org/health/diabetes-resources/the-vegan-diet-how-to-guide-for-diabetes

[8] By reducing cholesterol, a plant-based diet fights heart disease: http://www.pcrm.org/health/healthcare-professionals/nutrition-curriculum/section-one-preventing-and-reversing-heart-disease#veg_diets

[9] The cattle sector is a big driver of deforestation in the Amazon: http://www.greenpeace.org/international/en/publications/reports/slaughtering-the-amazon/

[10] 94% of American's think that animals in the agricultural sector should be free of cruelty: https://www.aspca.org/fight-cruelty/farm-animal-cruelty

[11] Research I helped conduct at Stanford University shows connections between adapting to climate change and reducing it, and the key lies in our behavior: https://www.garrisoninstitute.org/2013-09-16-15-39-23/cmb-video-presentations/cmb-video-2012/1318-marilyn-cornelius

[12] 2014 was the hottest year on record in 2015: http://www.climatecentral.org/news/2014-on-track-to-be-warmest-year-on-record-18041

[13] 2016 is the currently the warmest year on record: https://www.nasa.gov/press-release/nasa-noaa-data-show-2016-warmest-year-on-record-globally

[14] The World Health Organization has linked cancer to air pollution: http://www.cancer.org/cancer/news/world-health-organization-outdoor-air-pollution-causes-cancer

[15] Traffic pollution has been linked to birth defects: http://www.huffingtonpost.com/molly-rauch/traffic-pollution-linked-_b_3225148.html

[16] Animals in factory farms develop resistance to antibiotics, which in turn has been linked to 23,000 American deaths annually: https://awionline.org/content/human-health-impacts-factory-farming

[17] See the first essay in this chapter "Personal and Planetary Wellness: Win-Wins."

[18] Diabetes can be reversed with a raw vegan diet: http://www.treeoflifefoundation.org/service/reversing-diabetes-2/

[19] The China Study contains thousands of correlations between meat and dairy and lifestyle diseases. You can buy it on Amazon: http://www.amazon.com/The-China-Study-Comprehensive-Implications/dp/1932100660

[20] The World Peace Diet explores the connections between a plant-based diet and mechanisms for a more peaceful society. Find out more: http://www.worldpeacediet.com/

[21] Cowspiracy is a documentary that exposes the linkages between climate change and animal agriculture: http://www.cowspiracy.com/

[22] Forks Over Knives, a documentary, contains compelling evidence linking meat and dairy to lifestyle diseases such as obesity, diabetes, and heart diseases: http://www.forksoverknives.com/

[23] What the Health is a documentary about the connections between meat and dairy and lifestyle diseases, as well as corruption in the food industry: http://www.whatthehealthfilm.com/

[24] You can watch Earthlings for free on YouTube: https://www.youtube.com/watch?v=MwPNpy6TJf8

[25] Information about the 21-Day Vegan Kickstart program: http://pcrm.org/kickstartHome/

[26] Veganuary is about starting off the new year by going vegan: http://www.veganuary.com/

[27] For more information on Transcendental Meditation, which uses a single mantra to access deep consciousness, see http://www.tm.org/

[28] More information on Mindfulness-Based Stress Reduction, a system of meditation that is least religious: http://www.mindfullivingprograms.com/whatMBSR.php

[29] Vipassana is the technique taught by the Buddha, and consists of deep observation of the breath and bodily sensations. For more information, see: https://www.dhamma.org/en/about/vipassana

[30] Track the climate impact of your purchases using Oroeco: http://www.oroeco.com/

[31] Animal rights groups each have their own areas of focus, tactics, and culture. It would be good to explore them to see which approach suits you best.

[32] Learn more about Meatless Monday: http://www.meatlessmonday.com/

[33] This article is a simple summary of some of the principles in Native American cultures: http://themindunleashed.org/2015/02/native-american-code-ethics.html

[34] Read the report here: http://www.chathamhouse.org/publication/livestock-%E2%80%93-climate-change%E2%80%99s-forgotten-sector-global-public-opinion-meat-and-dairy

[35] The Alchemus Prime Diamond Model is built on the notion that an integrated approach to solving problems is more robust than working in silos. For more, see our website: http://www.alchemusprime.com/model/ and see Chapter 10.

[36] See Endnote #3 for this Chapter.

[37] This videos summarizes how meat and dairy are linked to the top 11 lifestyle diseases: https://www.youtube.com/watch?v=d0IhZ-R1O8g

Chapter 2: Eating For Wellness And Productivity

Cancerous Meat: Risk vs. Norm

The World Health Organization (WHO), via the International Agency for Research on Cancer (IARC), recently announced that processed meat, and probably red meat too, are linked to colorectal cancer.[38] The press release states:

"The experts concluded that each 50 gram portion of processed meat eaten daily increases the risk of colorectal cancer by 18%."

Yes, you read that correctly, 18%. We'll come back to this.

Needless to say, this caused uproar in social media circles. The Huffington Post published an article[39] stating that the WHO had clarified that people need not give up meat, but it would make sense to reduce processed meat consumption.

If you've read The China Study,[40] which lays out thousands of associations between meat and dairy products and cancer, diabetes, and other diseases, it's difficult not to pay close attention. Correlation is not causation, but with thousands of correlations, it might be worth changing our behavior from preventive and precautionary perspectives.

This is an important risk management problem. And we humans are usually pretty smart about it. I'll share with you what Steve Schneider,[41] the late climatologist and my esteemed mentor, used to ask during an invited talk, and the response he would always get (I'm paraphrasing):

Steve: How many of you have had a fire in your home?
(Maybe 1 or 2 hands go up in an audience of a few hundred people).
Steve: Okay, let's say 2%. How many of you have fire insurance?
(Almost all hands go up).
Steve: This is what risk management is about: being prepared for the likelihood of an event. We already do this, as you have just seen. We can apply the same logic to being prepared for climate change-induced events.

So, for all of us who have fire insurance but won't give up bacon, remember that the likelihood of a fire might be 2%, but, as I mentioned earlier, the increased risk of cancer with every 50 grams of meat eaten is

18%. I doubt we would want our children eating bacon and other processed and red meats, given that kind of risk…and if so, why would we eat it ourselves?

Giving up processed and red meats is also one of the most important actions we can take to address climate change, water shortages, deforestation, and famine.[42] The time to be smart and choose risk aversion (giving up meat) over a potentially dangerous norm (eating meat) is now.

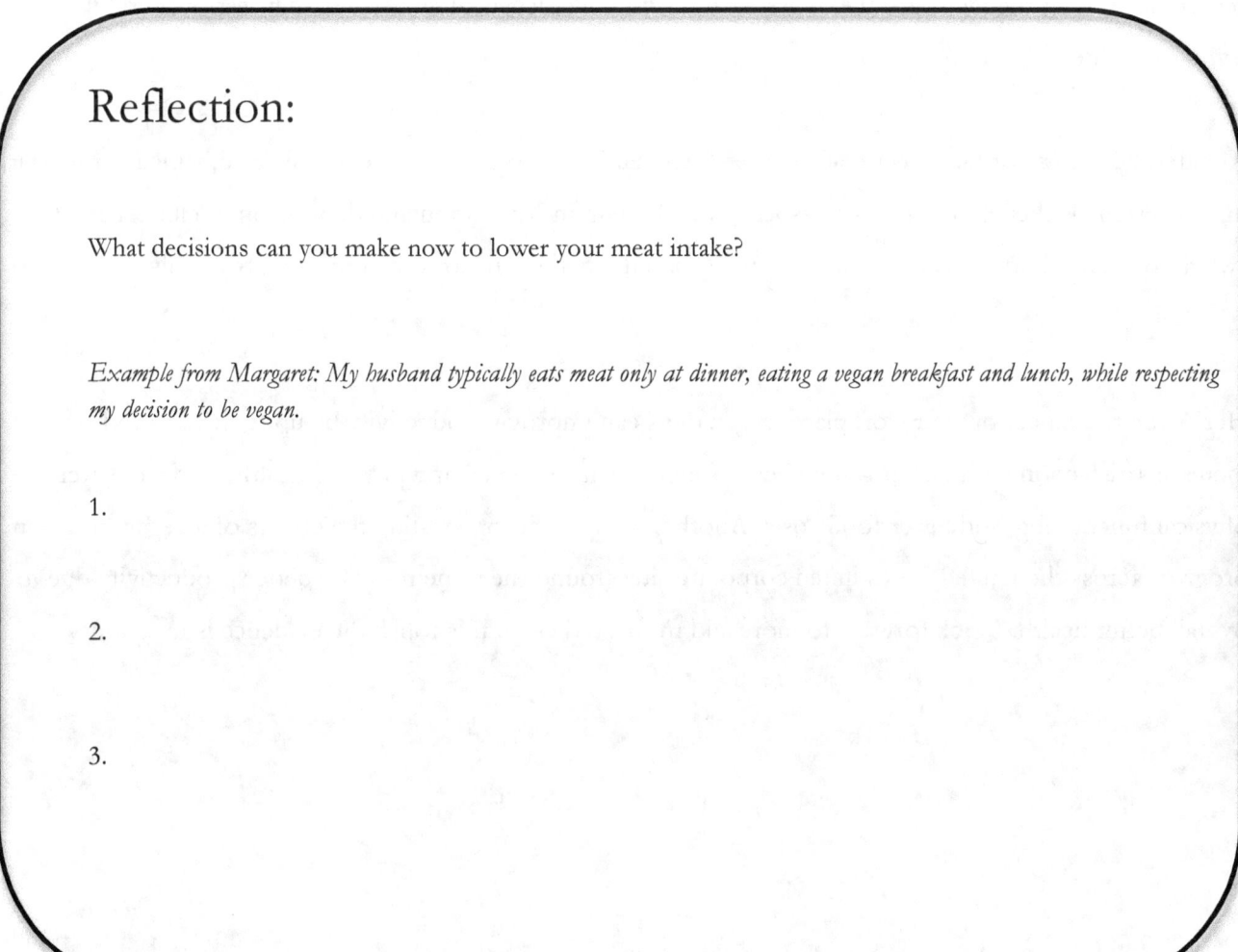

Reflection:

What decisions can you make now to lower your meat intake?

Example from Margaret: My husband typically eats meat only at dinner, eating a vegan breakfast and lunch, while respecting my decision to be vegan.

1.

2.

3.

Plant-Based Diet Improves Productivity

Peer-reviewed research indicates that plant-based diets can improve mood, reduce anxiety, and ameliorate fatigue.[43] And, contrary to the common belief that low-carb diets are important for weight loss, a study[44] comparing people with low-carb versus high-carb diets (with low levels of fat and protein) over a year indicates that the high-carb diet is associated with lower levels of anxiety and depression, and higher levels of wellness.

Apparently, the *type* of fat we eat really matters. Animal fats, in particular arachidonic acid, which is found in high levels in chicken and eggs,[45] are associated with brain inflammation and depression.[46] Yikes! Time to switch to peanut butter, avocado, or other plant-based fats that are conveniently and affordably available to you, folks.

Here's the relevance for your workplace: vegan diets can improve productivity by up to 46%,[47] AND: can increase satisfaction with diet, promote better sleep, enhance mental and physical health, provide better physical functioning, and lower food costs. Another study,[48] this one testing the effects of a vegan nutrition program across the United States in ten corporate sites, found the same results – better productivity due to overall better health. I look forward to more and more studies on this topic: the evidence is mounting.

Reflection:

When was the last time you noticed you were super productive? What did you eat?

Example from Marilyn: I wrote 4 blogs in 3 hours! I had a fruit smoothie with banana, pineapple, strawberries, blueberries, and vegan protein powder.

Meaty Meals: Get the Beef

I've written in the past about the myriad benefits of a plant-based diet, such as productivity,[49] climate stability,[50] and enhanced learning.[51] Let's look at the flip side of that: what eating meat actually does to the human body:

1. **Inflammation:** Eating red meat and saturated animal fat releases endotoxins,[52] a type of bacterium that causes inflammation in the body within an hour of eating the meal. Inflammation raises the risk of heart disease, stroke, and diabetes.

2. **Cholesterol:** Eggs, dairy, red meat, and especially liver, contain high levels of cholesterol.[53] Cholesterol also increases the risk for heart disease and stroke.

3. **Bad bacteria:** Red meat consumption causes a compound called trimethylamine-N-oxide (TMAO)[54] to be produced in the gut, and TMAO is linked to atherosclerosis[55] (hardening of the arteries), which can cause heart attacks.

4. **Aging:** Eating processed meat containing trans fats can shorten telomeres,[56] regions of our DNA. This shortening is associated with faster aging due to inflammation. Plant-based foods contain antioxidants and anti-inflammatory agents that slow down aging by lengthening telomeres.

5. **Diabetes:** Increasing meat intake by more than half a serving per day for four years raises the risk of type 2 diabetes by 48%, according to a series of studies.[57]

6. **Weight gain:** Eating more protein than we need can lead to weight gain,[58] because our body stores the extra protein as fat. Of course, weight gain can also result from eating too many carbs, and a host of other factors like hormonal imbalances, stress, medications, etc.

7. **Threat to planetary life-support system:** Eating animals destroys, pollutes, endangers, and/or destabilizes our land, water, fish, animals, wildlife, and climate, according to the National Commission on Industrial Farm Animal Production (NCIFAP).[59]

In sum, eating meat is a high price to pay for destroying our own well being, and that of other life forms. Now that you're sufficiently appalled, let's look at the silver lining: a plant-based diet basically does the opposite of the above, reducing inflammation and cholesterol, reducing harmful gut bacteria, slowing down the aging process, reducing the risk of type 2 diabetes and weight gain, and supporting a safe and healthy life now and in the future, for all. Bonuses include a lot more energy, and as I mentioned before, improved productivity. Everybody wins!

Reflection:

Which harmful effect of eating meat surprised you the most? Why?

Example from Marilyn: Aging — I didn't realize meat makes one age more! Yikes! I was startled to learn that it happens at the DNA level, proving that what we eat really matters.

Wellness at Work

At Alchemus Prime, we focus on fostering authentic leadership as a way to create optimal solutions for the earth's most pressing challenges, including climate change and lifestyle diseases. We work on wellness and climate change simultaneously[60] because they require similar behavior changes, and provide benefits for human health, climate stability, energy efficiency, animal welfare, water efficiency, land conservation, and much more.

Contrary to what you might expect, simply installing wellness programs isn't effective. In fact, research[61] suggests that wellness programs, including those with financial incentives, may not lead to tangible results[62] because they are perceived as yet another task, which may increase, not decrease, stress.

Research also indicates that employees prefer to be happy at work over a higher salary;[63] this is more evidence of the power of intrinsic motivators. Conversely, a harsh boss can lead to heart problems and more sick days for employees,[64] which harms productivity and the bottom line. A more effective approach, according to Stanford researcher Emma Seppälä, is to create a more humane, happy, and positive organizational culture,[65] which is exemplified by an organization's CEO. She states:

"So what leads to employee happiness? A workplace characterized by humanity. An organizational culture characterized by forgiveness, kindness, trust, respect, and inspiration. Hundreds of studies conducted by pioneers of positive organizational psychology, including Jane Dutton and Kim Cameron at the University of Michigan and Adam Grant at the University of Pennsylvania, demonstrate that a culture characterized by a positive work culture leads to improved employee loyalty, engagement, performance, creativity, and productivity. Given that about three-quarters of the U.S. workforce is disengaged at work — and the high cost of employee turnover — it's about time organizations start paying attention to the data."

Dr. Seppala encourages CEOs to do the following:[66]
1. Immediately show empathy e.g. through a personal phone call, when an employee has a crisis, thereby setting an example for positive social relationships, that satisfy employees' need for connectedness and belonging.
2. Foster a visible values-based culture, which ends up being more cost-effective than instituting fancy wellness programs.
3. Maintain a balance between showing humanity and not being "too nice" to ensure that accountability is also visible, and work gets done.

There are many dimensions to wellness, including how these dimensions relate to an organization's bottom line; the World Health Organization (WHO) captures them well in its framework,[67] highlighting the intricate connections between employee health and happiness and how well a business can function.

Workplace wellness is not only an integral part of Alchemus Prime's work, but also close to my heart. After working in a variety of settings, including academia, I intuitively felt there was a better system for treating human beings in the workplace. It is affirming to see research[68] clearly indicating that productivity and wellness depend on humanity in our professional environment. We are wired to connect, and will do our best work when we are part of a team that demonstrates, through its leadership and communication: genuine care, appreciation, respect, dignity, and kindness.

Reflection:

What could you do in your workplace to improve wellness?

Example from Marilyn: Have clear and timely communication – so essential!

Ditch Diets and Reject Resolutions for Wellness

Every year, thousands of people make New Year's resolutions. Most times it is to do with health and wellness. I am focusing on food-related resolutions, as a case in point. Reports show that diet-related resolutions don't work. For example, those who do go on a crash diet,[69] lose some weight and then stop, usually put on more weight by going back to their old habits.

Well, it may be time to try a different approach. Taking one step at a time and turning a new behavior into a habit is a useful approach. For instance, if someone wants to avoid diabetes, then a small first step might be to give up sugary soda. Then comes the longer-term commitment: following through. Changing habits[70] can be difficult, but starting is half the battle.

Let's go a step further. If we are genuinely interested in changing our food-related habits to avoid diabetes,[71] for example, the trick is to change gradually without feeling that it's a burden or we are being deprived. Once we've stopped the soda and that has become a new healthy habit, we can add on the next behavior change. Perhaps this is skipping dessert when you go out to restaurants. You don't have to miss all the fun though: you could substitute refined sugars with fruits in home made desserts.

Over time, you'll find that you feel much better, and then you'll want to give up sugar altogether. Sugar is, according to research,[72] more addictive than heroin; so don't give yourself a hard time, just stay on the path to wellness.

Some other not-so-healthy habits, and addictions include: taking refined, processed and animal products that typically contain addictive flavor chemicals, antibiotics, and other nasties; smoking and drug use; alcohol abuse; and being physically inactive. There's also depression, which research suggests is caused by a lack of social connection,[73] can be clinical or non-clinical. According to Harvard Health,[74] depression can also be caused by the way the brain is able to regulate emotion, our genes, stress, seasons, medication, and trauma.

Once we master the art of changing an existing behavior into new habit, we gain confidence and feel encouraged. We may even decide to embrace wellness as a lifestyle.[75] Simple reflection on our own lives can, along the way, give us the insight needed to improve, and sustain the improvement.

Habits are formed over many years and it feels very comfortable to do things the same way every time. However, to get a different result, the practices or habits need to change. This quote makes a lot of sense in this context:

"Insanity: doing the same thing over and over again and expecting different results." – Author Unknown

For 2017 and beyond, try the approach that actually works. Ditch diets and reject resolutions. Focus on building one healthy habit at a time, and surround yourself with supportive people, starting with your number one advocate and best friend: YOU.

This essay is written by the ever-energetic Margaret Cornelius, (an)altruist at Alchemus Prime.

Reflection:

What is the one simple action you can change starting today to be healthier?

Example from Marilyn: Replacing cow's milk with almond, soy, rice, or coconut milk.

Lifestyle Diseases: Prevention is the Cure

There's so much written about prevention of lifestyle diseases,[76] but the statistics[77] are still showing alarming rates of diabetes, high blood pressure, cancers, heart disease, strokes, and other diseases around the world. These ailments usually affect adults and cause premature deaths. However, with unhealthier lifestyle habits like poor diets, drug and alcohol use, and lack of exercise, younger people are now suffering as well.

There are multiple lifestyle risk factors, which predispose us to these diseases but that most of us don't understand clearly. These risk factors[78] are classified as modifiable or non-modifiable. Our advancing age, gender and inherited genes are mostly non-modifiable. However, lifestyle habits are definitely modifiable. Even though habits maybe formed early in life, we can certainly change these given the right motivation and support. Clearly understanding the harm in these risk factors can motivate us to change.

Major modifiable lifestyle risk factors include:

- Unhealthy diet;
- Lack of physical activity;
- Unhealthy weight;
- High stress;
- Alcohol and drug abuse;
- Smoking cigarettes, and exposure to tobacco smoke.

Another issue is that many of these lifestyle diseases creep up on us without any warning signs or symptoms. The typical symptoms of diabetes[79] may be vague or absent in adults, such as frequent peeing, or feeling thirsty and tired all the time. High blood pressure[80] usually does not have any warning sign or symptoms. Cancers may also have no or vague symptoms,[81] depending on the type, where they occur in the body, and how advanced they are.

Usually, there is too much information, or (mis)information in the public domain regarding many of these risk factors and the warning signs of the lifestyle diseases. Clever advertising convincingly bombards us into believing that unhealthy habits are cool. It doesn't help that cheaper and easily available unhealthy choices are everywhere, such as processed, packaged and fast foods; sugary beverages; cigarettes and tobacco products; and a variety of drugs.

Rather than waiting for symptoms to occur, which means it's going to take much longer to reverse your disease, it's better to get rid of lifestyle risk factors and build healthy habits, as I mention earlier in this chapter. Learning about risk factors and starting to address them as early as possible will save our loved ones and us all kinds of aches later in life. If you are an optimistic person like me, get a head start on living a life of wellness. Doing things right for ourselves makes us feel good, and we can become role models for loved ones, save on health care costs in the long run, and be free to enjoy a productive and fulfilling life.

I would encourage you to take stock of where you are on the spectrum of wellness and lifestyle diseases. Find out how you're doing with regard to all these risk factors via a biometric screening, including blood tests, and start taking a preventive approach.

This essay is written by the ever-energetic Margaret Cornelius, (an)altruist at Alchemus Prime.

Reflection:

What risk factors do you have that you might start to reverse now?

Example from Margaret: I don't smoke, and avoid passive smoking to minimize my risk of cancer.

Plant-Based Food: A Doctor's Personal Journey

I have been a meat eater for more than sixty years.

Despite my forty-year career in medicine, I was rarely exposed to plant-based medicine[82] as a therapy for lifestyle diseases. I supported nutritionists in their work, but with a view to solving medical conditions and not preventing illnesses. When I specialized in diabetes management and control, I educated myself on good nutrition so I could give advice to my patients on what foods to take to help control diabetes, and related lifestyle conditions such as obesity, high blood pressure, high cholesterol and heart problems. Knowing more about nutrition, however, did not make me change my eating habits much, except for stopping sugar and sugary foods and avoiding red meats. Luckily, I was not addicted to fatty, fried or fast foods.

Then, something drastic happened.

My daughter, Marilyn, decided at age 16 to become vegetarian. I panicked. Where will she get her protein? What am I going to feed her? What if she gets sick? My medical mind was in chaos. I was brainwashed as a child and while growing up in school, that protein comes from meat and dairy products. My medical training confirmed meat and dairy as major sources of protein. However, when I researched further, I was somewhat appeased that Marilyn would be 'okay' if she took dairy products and eggs.

Many years later, another panic attack came about when Marilyn, now grown up and very independent, announced that she was going vegan. My worries increased, but at the same time, I was having health issues that seemed to be related to food intake. My colleague, who is a raw food advocate and vegan, advised me to stop all meat and seafood, and to eat raw food. I was shocked and had difficulty trying to make this drastic change. Then my daughter came home for a holiday, and gave me The China Study[83] to read. I began reading it slowly, but remained skeptical.

During her yearly trips to Fiji, Marilyn was very creative with vegan meals and was cooking delicious food happily and inexpensively. For her, vegan dishes were easy to prepare, as she was adapting her vegetarian favorites to vegan versions. Vegetables are readily available and very affordable compared to meat in Fiji. Over time, I realized fear was preventing me from changing my diet: fear of becoming ill, not having enough protein, and being ostracized by my family and friends.

Soon after Marilyn left for California after a three-month stay with me in early 2015, I became vegan. Later that same year, Marilyn wrote a vegan recipe book called 'Food of Love,'[84] containing easy-to-make and nutritious meals. She shared this with me to help me honor my decision. Now, I feel very confident that I made the right choice and am continuing to make better choices.

Since visiting her in California in 2017, I have also become gluten-free (she made that decision at the end of 2015), and am feeling very healthy and alert. I taught myself to make corn tortillas to replace our traditional "roti" or wheat bread. I started walking more than 3 miles daily, with unprecedented energy levels. Despite my fear and brainwashing regarding animal protein, I am a very happy and healthy vegan.

There is a group of physicians called the Physicians Committee for Responsible Medicine (PCRM)[85] that advocates for healthy nutrition practices to manage lifestyle diseases. For me, they are much more informative than some disease-related websites. My risks for lifestyle diseases have disappeared completely within a year, and I am able to maintain a healthy weight easily, with an abundant diet. Where are you in the diet journey?

This essay is written by the ever-energetic Margaret Cornelius, (an)altruist at Alchemus Prime.

Reflection:

What is one fear that holds you back from changing your diet to a plant-based one?

Example from Margaret: For me, it was what people would think and say about me.

Plant-Based Medicine: A New Era

Hippocrates said, *"Let food be thy medicine, and medicine be thy food."*[86]

This notion is more relevant today than ever before and needs greater focus due to escalating risk factors[87] for nutrition–related diseases. Obesity, diabetes, and heart disease are but the tip of the iceberg for nutrition and lifestyle-related diseases.[88]

During my medical school education in the 1970s in Fiji, nutrition was an optional subject. Only a few hours were spent on nutrition education, and students and teachers alike considered it unimportant. More attention was paid to teaching and learning anatomy, physiology, biochemistry, pharmacology, and other big 'ologies' than nutrition, even though lifestyle and nutrition-related diseases were already common.

In those days, patients listened more to doctors than allied health workers, such as physical therapists and nutritionists. Consulting a nutritionist was not a priority and doctors didn't give detailed advice on nutrition and healthy cooking practices due to their own lack of training and the attitude that nutrition was a minor aspect of treatment. Patients also wanted to believe that the drugs prescribed by doctors were a "quick-fix" for their ailments.

I was one of these uninformed physicians, but fortunately, not for long. When I specialized in diabetes care in 1996, my patients kept asking me about specifics of diabetes control through daily dietary changes. I realized a gap in the National Diabetes Center's approach to diabetes management. In response, I co-produced a 44-page book called 'Food and Diabetes' for people with diabetes; this book contained specific guidelines on how to reduce dietary sugar intake. For example, skipping refined sugar in tea and coffee, and avoiding sugary soda and juices.

About a decade later, my medical and nursing colleagues and I developed a more holistic approach. We designed and deployed a Green Prescription, which prescribed lifestyle changes, but *not* drugs. We built on New Zealand's Green Prescription, which promoted physical activity.[89] Our Green Prescription included basic tips on good nutrition, exercise, responsible use of alcohol, avoidance of tobacco products and drugs, and stress relief for all individuals who were deemed to be at risk of diabetes.

Recently, after frequent trips to the U.S., I realized that the situation has been similar in medical schools here. Studies[90] show that most U.S. physicians are inadequately prepared to give useful nutrition advice to their patients.

However, the good news is that some medical training providers are realizing the benefits of treating lifestyle diseases with healthy eating habits, and are teaching medical students nutrition and healthy cooking methods.[91] Over the years, the benefits of whole and natural plant-based foods,[92] consumed in their natural forms, are becoming clearer and more acceptable.

The Plantrician Project,[93] for instance, defines a "plantrician" as a physician who understands the role of plant-based foods in health and wellness; these physicians teach nutrition and healthy food preparation to medical students. There is also a group of 12,000 physicians called the Physicians Committee for Responsible Medicine (PCRM)[94] that advocates for healthy plant-based nutrition practices to manage nutrition and lifestyle related diseases. The PCRM is a great resource for finding out how plant-based foods can optimize our health and wellness.

Now that the tide is turning, we as consumers and good citizens must be proactive and challenge our health care providers to gather and share knowledge about plant-based medicine. We can help accelerate the shift away from harmful pharmaceutical drugs that have many side effects and interactions with other drugs.

So, based on what I've shared about this shift toward plant-based medicine, it seems the time has come to seriously consider letting "food be thy medicine" instead of drugs.

This essay is written by the ever-energetic Margaret Cornelius, (an)altruist at Alchemus Prime.

Reflection:

List three foods or herbs that are medicinal for you and why.

Example from Marilyn: Apple cider vinegar, because it keeps my pH alkaline; ginger, which keeps me warm, and lemon for immune system protection.

1.

2.

3.

Chapter 2 Notes

---

[38] Read the International Agency for Research on Cancer's (IARC) monograph here: https://www.iarc.fr/en/media-centre/pr/2015/pdfs/pr240_E.pdf

[39] http://www.huffingtonpost.ca/2015/10/30/who-bacon-cancer_n_8432392.html

[40] The China Study contains thousands of correlations between meat and dairy and lifestyle diseases. You can buy it on Amazon: http://www.amazon.com/The-China-Study-Comprehensive-Implications/dp/1932100660

[41] To read more about the late Steve Schneider's work and legacy, see his primary website: http://stephenschneider.stanford.edu/

[42] See Chapter 1 for the essay entitled, "Going to Hell in a Handbasket or Converging on a Solution for more on how meat production damages the environment.

[43] For the peer-reviewed article in *Nutrition Journal*, see: https://www.ncbi.nlm.nih.gov/pmc/articles/PMC3293760/

[44] Low-fat diets have a better effect on mood than low-carb diets: https://www.ncbi.nlm.nih.gov/pubmed/19901139

[45] Chicken and eggs contain high levels of a harmful fatty acid called arachidonic acid, according to the National Cancer Institute: https://epi.grants.cancer.gov/diet/foodsources/fatty_acids/table4.html

[46] The presence of animal fats can harm the brain, leading to inflammation and depression: https://www.researchgate.net/profile/Tahira_Farooqui/publication/6547424_Modulation_of_inflammation_in_brain_A_matter_of_fat/links/09e415017194a18830000000/Modulation-of-inflammation-in-brain-A-matter-of-fat.pdf

[47] Vegan diets can increase productivity by up to 46% according to an article in the *Annals of Nutrition and Metabolism*: http://www.pcrm.org/sites/default/files/pdfs/health/medstudents/A%20Worksite%20Vegan%20Nutrition%20Program%20Is%20Well-Accepted%20and%20Improves%20Health-Related.pdf

[48] An article in the American Journal of Health Promotion found that a plant-based diet improves productivity, and lowers depression and anxiety: https://www.ncbi.nlm.nih.gov/pubmed/24524383

[49] See the essay entitled "Plant-Based Diet Improves Productivity" in this Chapter.

[50] See Chapter 3, especially the essay entitled "Research: Vegan Diet Crucial to Address Climate Change."

[51] See my blog on how children feel and learn better on a plant-based diet:

http://www.integritusprime.com/veg-kids-feel-and-learn-better/

[52] How endotoxins in meat cause inflammation: https://nutritionfacts.org/2012/09/20/why-meat-causes-inflammation/

[53] The University of California San Francisco (UCSF) Medical Center provides information on the cholesterol content of foods: https://www.ucsfhealth.org/education/cholesterol_content_of_foods/

[54] The National Institutes of Health explain how gut microbes break down meat into compounds that can cause heart disease: https://www.nih.gov/news-events/nih-research-matters/red-meat-heart-disease-link-involves-gut-microbes

[55] Read more about atherosclerosis: https://www.nhlbi.nih.gov/health/health-topics/topics/atherosclerosis

[56] How processed meat consumption is potentially shortening our lifespan: http://ajcn.nutrition.org/content/88/5/1405.full

[57] Red meat intake, when increased over time, is associated with higher risk of type 2 diabetes: http://jamanetwork.com/journals/jamainternalmedicine/fullarticle/1697785

[58] Eating too much protein can contribute to weight gain, according to the National Center for Health Statistics: https://nchstats.com/2010/03/03/adults%E2%80%99-daily-protein-intake-much-more-than-recommended/

[59] Download the full report here: https://www.ncifap.org/reports/

[60] To learn more about how we work, see: http://www.alchemusprime.com/how-we-work/

[61] A good boss could be even more important than a wellness program for the wellness of workers: https://hbr.org/2016/04/good-bosses-create-more-wellness-than-wellness-plans-do

[62] Download this guide to see which metrics matter for employee health management: http://populationhealthalliance.org/publications/program-measurement-evaluation-guide-core-metrics-for-employee-health-management.html

[63] Read more about why employees prefer to be happy rather than have a higher salary in these articles: http://fortune.com/2016/09/14/corporate-social-responsibility-top-talent/?mkt_tok=eyJpIjoiTkRNd1lXVXhZbVV6TkRRMSIsInQiOiI3SEZNMkpKZFhUdDd4ZTlObFJlYmFRQUE4N2pHOXFsZTFFTMGhpdUsrWlZrdWg4OElTRm9JY0JYY0pWdHVUSE1TVXRhT3hjZU1xd1BDNFwvZ3ZSdXd3WFNpRlh1UUZHaUFzcU9aRlVyTDNsZUMwK3l4ckxvYkZ1WWFoNEZUMjdoMnUifQ and https://www.aat.org.uk/about-aat/press-releases/britains-workers-value-companionship-recognition-over-big-salary

[64] Poor leadership can lead to health problems in employees: http://ki.se/en/news/poor-leadership-poses-a-health-risk-at-work

[65] A humane workforce culture leads to employee happiness: http://www.emmaseppala.com/best-ceos-earth-better/?utm_content=buffer32e8d&utm_medium=social&utm_source=twitter.com&utm_campaign=buffer

[66] Ibid.

[67] Download the WHO Healthy Workplace Framework and Model here: http://www.who.int/occupational_health/healthy_workplace_framework.pdf

[68] See essay entitled: "McKinsey: Four key Leadership Behaviors" in Chapter 10

[69] Why crash diets fail: http://straighthealth.com/pages/articles/darticles/crash-diets-dont-work.html

[70] Read more about how to change habits: https://psychcentral.com/lib/7-steps-to-changing-a-bad-habit/

[71] How to prevent, reverse, or manage diabetes using diet: http://www.pcrm.org/health/diabetes-resources

[72] The disturbingly addictive properties of sugar: https://www.forbes.com/sites/jacobsullum/2013/10/16/research-shows-cocaine-and-heroin-are-less-addictive-than-oreos/#534012372427

[73] How social connection can help with depression: https://psychcentral.com/news/2014/03/20/social-connections-can-help-to-reduce-depression/67371.html

[74] The causes of depression, according to Harvard Health: http://www.health.harvard.edu/mind-and-mood/what-causes-depression

[75] For more on wellness as a lifestyle choice, read Margaret's blog post here: http://www.azentive.com/2017/01/07/well-workforce-your-lifestyle/

[76] Lifestyle diseases are defined as illnesses that relate to the way people live their lives: http://www.medicinenet.com/script/main/art.asp?articlekey=38316

[77] Read the overview for the World Health Organization's Global Status Report on Non-Communicable Diseases: http://www.who.int/nmh/publications/ncd-status-report-2014/en/

[78] The World Health Organization has a list of risk factors you can read about here: http://www.who.int/gho/ncd/risk_factors/en/

[79] For more information on typical symptoms of diabetes, see the American Diabetes Association's list: http://www.diabetes.org/diabetes-basics/symptoms/?referrer=https://www.google.com/

[80] High blood pressure can be without symptoms, making it hard to detect: http://www.heart.org/HEARTORG/Conditions/HighBloodPressure/UnderstandSymptomsRisks/What-are-the-Symptoms-of-High-Blood-Pressure_UCM_301871_Article.jsp#.WULWRxMrKcY

[81] Cancer has vague symptoms, or none at all: http://www.cancercenter.com/terms/cancer-symptoms/?source=GGLPS01&channel=paid+search&invsrc=Non_Branded_Paid_Search_Google_General_Search&utm_device=c&utm_budget=Corporate&utm_site=GOOGLE&utm_campaign=Non+Brand%3ETop+Terms&utm_adgroup=Symptoms%3ECancer+Symptoms%3EBMM&utm_term=%2Bsymptoms+%2Bof+%2Bcancer&utm_matchtype=b&k_clickid=181c8acc-fd45-4a1a-a211-972b196cdead&k_profid=422&k_kwid=3975743

[82] See also, "Plant-Based Medicine: A New Era," also in Chapter 2

[83] The China Study contains thousands of correlations between meat and dairy and lifestyle diseases: http://www.amazon.com/The-China-Study-Comprehensive-Implications/dp/1932100660

[84] For more about Food of Love, see our website: http://www.alchemusprime.com/our-books/

[85] Read more about the Physician's Committee for Responsible Medicine (PCRM) on their website: http://www.pcrm.org/about/about/about-pcrm

[86] For more quotes by Hippocrates, see: https://www.goodreads.com/author/quotes/248774.Hippocrates

[87] See essay in this chapter, entitled "Lifestyle Diseases: Prevention is the Cure" for more on risk factors.

[88] For a list of nutrition-related diseases, see the World Health Organization's technical report summary: http://www.who.int/dietphysicalactivity/publications/trs916/summary/en/

[89] To learn more about New Zealand's Green Prescription, see: http://www.health.govt.nz/our-work/preventative-health-wellness/physical-activity/green-prescriptions

[90] This peer-reviewed journal provides more information on the role of nutrition in medicine: https://www.ncbi.nlm.nih.gov/pmc/articles/PMC4594871/

[91] Read about how a medical school in New Orleans, Louisiana, is providing cooking classes to medical students: http://www.cbsnews.com/news/scalpel-or-spatula-some-medical-students-have-new-required-cooking-course/?ftag=CNM-00-10aab8a&linkId=32951450

[92] Benefits of switching to a plant-based diet include blood sugar control, better digestion, more energy, and less pain: http://www.onegreenplanet.org/natural-health/quick-benefits-youll-see-by-switching-to-whole-foods-plant-based-diet/

[93] Learn about the Plantrician Project: http://plantricianproject.org/

[94] See note #85.

Chapter 3: Eating Your Way To A Stable Climate

Business as Usual = Catastrophic Food Shortages

A scientific model developed by the Anglia Ruskin University[95] indicates that without substantial change in behavior and climate policy, the world will suffer drastic food shortages and face a food system collapse by 2040. Global food supply issues, climate change, as well as other related issues such as deforestation, land degradation, water scarcity, poverty, and famine,[96] are related to unsustainable industrialized animal agriculture, which accounts for as much as 51% of global greenhouse gas emissions.[97]

A recent article from scientist Arthur Biglan[98] supports what I've been saying all along and what Alchemus Prime is founded upon: behavioral sciences. Our behavior is crucial to human well being. We humans *are* the solution to our health and the planet's.[99] According to Biglan, and confirmed by research I'm familiar with at Stanford University and elsewhere, behavioral science has had success in obesity prevention,[100] sanitation,[101] and other well being-related issues.

It makes sense to apply behavioral science to successfully address the intersecting problems of human wellness and climate change.[102] The good news is: we're already doing it! Research-based behavioral energy efficiency programs are going mainstream,[103] and behavior change can help double world food supply.[104] We've got this. We must act, because every action matters.

Reflection:

How might you reduce food waste in your home or workplace?

Example from Margaret: Use leftovers creatively instead of throwing them away.

1.

2.

3.

Food of Love: Recipes for Life

Think of a new X-Men character that enhances every meal into a plant-based meal. Then realize: that character is you.

Now, what does this have to do with anything, much less my first cookbook, you might ask.

Well, this:

Over a year ago, I wrote a guest blog for Asking Nature,[105] the Biomimicry Institute's official blog. In that piece, I shared why a transition to a plant-based diet is a life-affirming and survival-oriented mutant behavior. Yes, Wolverine, we're talking about regeneration – of forests.[106] Plant-based living allows humans, animals and entire ecosystems to thrive together, while addressing the climate, drought, food insecurity, land degradation, extinction, lifestyle disease and other related challenges.[107]

Plant-based living is the strongest win-win solution we at Alchemus Prime can recommend, based on the science that is currently available. According to the BBC, the Governator also recently recommended we all become at least part-time vegetarians to protect the planet,[108] so he is in agreement with the science too.

I'm not making this stuff up. What I *have* made up, though, is a cookbook filled with recipes from my own experience that are biomimetic in the sense of drawing inspiration from nature and aligning ingredients with Life's Principles[109] in at least the following ways:

1. **Adapting to Changing Conditions:** these recipes are designed to be adaptive to the resources you have, including a list of variations that make them easy to improvise and change based on changing circumstances, while staying within the parameters of being earth-friendly;
2. **Evolving to Survive:** the ingredients have evolved, with my help, to use plant-based ingredients that are more aligned with nurturing all life and reducing greenhouse gas emissions;
3. **Becoming Resource Efficient:** Plant-based foods are much more efficient[110] to produce than animal-based foods, so it makes sense to use fruits, vegetables, grains, nuts, and legumes to satisfy our appetites for food and, as it turns out, for our health[111] too.

Don't take my word for it. See for yourself. It's called *Food of Love*, and it's available via the Alchemus Prime website.[112] For a short description of the book, see the Further Reading section in this volume.

Reflection:

How do you connect with nature through food? Draw your response.

Climate Change: What Would Nature Do?*

Recently I visited Xunantunich, a site where Mayan ruins are being uncovered in the impoverished nation of Belize, formerly British Honduras. Aside from the sheer beauty and powerful energy of the site, a pair of trees our guide pointed out struck me. One was called the Poisonwood Tree, (*Metopium brownei)*[113] because of the strong allergen in its bark. The other, a reddish tree, which I later found out is called the Gumbo-Limbo tree (*Bursera simaruba)*,[114] contained the antidote.

From a behavioral science perspective, which underpins most of my personal and planetary work, this phenomenon of adjacent peril and cure would warrant careful observation to inform intervention design principles that would:

a) optimize for desired behaviors (avoiding poisonous plant altogether, and if encountering it, knowing about the antidote and obtaining it immediately, as well as sharing the learning with others), and

b) minimize undesirable behaviors (encountering and ingesting the poisonous plant without knowing about it).

Of course, variables such as time taken for poisons and antidotes to take effect, and potency of each, can be precisely studied in the lab. When it comes to climate change, though, as my late mentor Steve Schneider[115] used to say, we don't have "Lab Earth" on which to try out our experiments. How, then, might we address climate change using our everyday actions?

We can turn to biomimicry to help us answer this behavioral question. Climate change is upsetting the balance to the global atmospheric system. When Ma Earth faces an imbalance, as climate change presents, nature produces her own experiments: mutations. Mutations[116] are permanent genetic changes that produce different characteristics in individuals of a species. Some mutations result in failure and death in the environment they enter; others serve to pioneer new adaptive strategies for survival and success of a species.

In order to address climate change, we must experiment with *mutations of our behavior* to achieve win-win solutions, that is, to bring our behavior back into alignment with all life and ensure no harm is being done to our precious earth for which we have no duplicate version. Discerning win-win solutions is not difficult if we take into account impacts of our actions on various aspects of the context in which we live. For example, if we "mutate" our eating behavior, say by eating plant-based foods when social norms support meat and dairy, what impact does that have on our staple meat-centric diets?

Why focus on diet, anyway? Why challenge the norm of meat-eating?

An announcement[117] from the World Health Organization (WHO) that red and processed meat intake is linked to colorectal cancer has caused quite a stir in the media. Worse, we already know that producing red meat is extremely inefficient and has many other harmful impacts: animal agriculture has the highest greenhouse gas[118] impact compared to transportation and other sectors, requires copious amounts of grain as feed, and 100 times the water[119] compared to growing vegetables and grains, and requires the slaughter of billions of farmed animals worldwide annually.[120] Plus, we are losing rainforests at the rate of an acre per second,[121] with 70% of the Amazon rainforest now destroyed due to beef production.[122]

Research tells us that a vegan diet has the lowest greenhouse gas footprint,[123] lowest water footprint,[124] avoids cruelty to animals,[125] can reverse heart disease,[126] obesity,[127] and type 2 diabetes,[128] can reduce colon cancer risk,[129] and requires less land compared to meat and dairy.[130] Eating a plant-based diet is, by definition, a win-win solution, and has garnered support by the United Nations, as evidenced by both a report[131] from the United Nations Environment Program and the somewhat famous FAO's *Livestock's Long Shadow* report.[132]

The obstacle then, is how to increase acceptability of this mutant behavior. How might we rapidly shift societies that are accustomed to meat and dairy, to plant-based foods that are better for their health and the well being of the planet?

The answer integrates behavior change, design thinking, biomimicry, and meditation:

Behavior change: If humans change their dietary habits and become plant-eaters, we will be able to free up grasslands and pasturelands currently used to graze livestock, and convert them to native forests to sequester more carbon than we have added to the atmosphere since the industrial era.[133] Plus, this can improve human health, reversing the global epidemics of diabetes and obesity. From a moral standpoint, we will also be able to spare animals from being brutally treated in industrialized factory farms, and regenerate a healthy planet for our children and theirs.

Design thinking: There are many creative pathways (including biomimicry) for how to initiate and implement a shift to a plant-based diet, which human-centered design can be employed to help discover and prototype iteratively to find the best implementation approaches.

Biomimicry: If we humble ourselves and look to nature as mentor, we find that a given piece of land can support more herbivores than carnivores. This is because herbivores are eating directly from the land, with a much more efficient conversion of energy from plants than carnivores, who convert energy from the bodies of herbivores. Translating this to our animal agriculture system, we see we have greatly skewed this balance,[134] and that it is tremendously inefficient to produce meat compared to plant-based foods.[135] As we perpetuate this system, rooted in deforestation, biodiversity extinctions continue.[136] Nature promotes all life by adapting to change, evolving to survive, and being resource efficient; in order to follow these Life's Principles,[137] we must let nature lead.

Meditation: Each person can become a leader in this shift, but an important prerequisite is to connect deeply with self, family, colleagues, community, and planet. In the process of connecting deeply, we become observers of our thoughts, behaviors, emotions, and those of others; this helps us be less reactive, and more calm. This level of connection and self-observation requires and can be expedited through a regular meditation practice.

These four pillars form the science-based Alchemus Prime Diamond Model[138] for leadership and human innovation, inspired by nature.

The answer, then, to "climate change: what would nature do?" is: nature would adapt. Adaptation is a behavioral endeavor. The antidote (plant-based eating) is right next to the poison (meat and dairy). The distance between the two, and how you can participate in the loving movement that nurtures all life? Your daily behavior.

Marilyn originally wrote this as a guest blog for Asking Nature, The Biomimicry Institute's official blog. The essay has been modified slightly for this book.

Reflection:

How would you encourage a loved one to shift to a plant-based diet?

Example from Marilyn: For my great-aunt, I do it by cooking her tasty vegan meals.

Woof! Your Pet's Carbon Footprint

I'm a dog fanatic. When I see a dog walking his or her human(s), I say hello to the pooch, not the human! It's terribly rude; I keep failing at changing this behavior. I simply adore dogs!

That's why this piece was tough to write. It's about the carbon footprint of our beloved furry family members. There are many dimensions to this discussion, and in true Alchemus Prime fashion,[139] they relate to the deep interconnected nature of climate change and wellness. Let's examine each of the dimensions of the issue of "carbon paw prints:"

Meat & Land

Dogs (and cats) are carnivores, and much of their impact stems from this fact. Here's how an article[140] in Salon sums it up:

"An average-sized dog consumes about 360 pounds of meat in a year and about 210 pounds of cereal. Taking into account the amount of land it takes to generate that amount of food and the energy used, that makes your dog quite the carbon hound. A 2009 study by New Zealand's Victoria University of Wellington concluded that pet dogs have carbon paw prints double that of a typical SUV. John Barrett of the Stockholm Environment Institute, in York, Great Britain, confirmed the results of the New Zealand study. "Owning a dog really is quite an extravagance, mainly because of the carbon footprint of meat," Barrett told New Scientist Magazine."

Cars

Another way of looking at the impact of our pets is to compare them to vehicles. Authors Robert and Brenda Vale did the math[141] and apparently:

"A medium-sized dog has the same impact as a Toyota Land Cruiser driven 6,000 miles a year, while a cat is equivalent to a Volkswagen Golf."

Of course, some of us buy SUVs to transport our pooches, so that's double trouble for the climate.

I'm a vegan who doesn't drive, y'all. And, I dream about having two medium or large dogs someday. But, as a climate change professional, I don't know how to fulfill my dream AND maintain my integrity.

Consumerism

It turns out that having a dog is expensive, and can become an extension of a materialistic lifestyle, with even more STUFF[142] – plastic bags, toys, processed foodstuffs, and so on. The news is not good:

"Pet product sales are expected to grow to $95 billion by 2017. Only a few corporations control 80 percent of the worldwide market for pet food, companies like Mars, Proctor and Gamble, and Colgate… Starting with the meat that constitutes most of the product itself, to the materials used to produce and package the food, and ending with the trucks that bring the products to market, all are enormously damaging to the planet. Fossil fuel is used all along the production line and carbon pollution is significant, as well as pollution from fertilizers, pesticides and the final product, animal waste."

Waste

It gets worse. Dog poop[143] is harmful to humans, waterways, and the planet in general:

"In the U.S…. dogs [cause] 10 million tons of waste a year. In 1991, the Environmental Protection Agency placed canine feces in the same category of pollutant as oil, herbicides, insecticides, and other deadly contaminants. Dog waste is full of bacteria that is harmful to humans, causing cramps, diarrhea, kidney problems, and many other intestinal illnesses. According to the EPA, waste from just 100 dogs could produce enough bacteria in three days to close a bay and all watersheds within 20 miles to all shellfish fishing and swimming."

Cat poop is even worse:[144] it can lead to severe harm to humans and other species through a parasite known as toxoplasma gondii.[145]

Biodiversity

Dogs and cats can threaten endangered species. According to the Science Encyclopedia:[146]

"In many places, vulnerable native species have been decimated by non-native species imported by humans. Predators like domestic cats and dogs, herbivores like cattle and sheep, diseases, and broadly-feeding omnivores like pigs have killed, starved, and generally outcompeted native species after introduction."

Wellness

There are two aspects to wellness I want to address. The first is very obvious: our own wellbeing. Pets improve our lovey-dovey feelings and social bonding by stimulating the release of oxytocin;[147] they reduce our stress, and help with our loneliness with their unconditional love and companionship. The prospect of living without that, as I do now, is not welcome to me.

But the second aspect of wellness is even more important to me: pets' well being. When we domesticate dogs and cats, we "train" them, which seems to me a euphemism for controlling them and punishing them until they lose much of their natural predatory and survival instincts. Then, they can live with us and be on our leashes. What about their natural tendencies, instincts, and desires to be free, to roam, to hunt, and to simply be themselves?

At Alchemus Prime we help our clients find and hone their true selves.[148] I feel very uncomfortable that we so often take that true nature away from animals and subject them to our lifestyles and whims. If anything, this would be the main reason that I would never own a pet again. Even that word, "own" is repugnant to me. I would never want to be owned by anyone.

So, my personal struggle continues, no doubt, but there are solutions[149] that we can consider, especially if we already have and love our pets:

"The choices we make as pet owners can reduce our pets' carbon paw print. Beef production has a much higher environmental impact than chicken or fish production, so by avoiding pet foods (and human foods for that matter) derived from beef, we help reduce carbon pollution. Quit flushing the cat litter. Switch from clay litter to a more sustainable litter that is biodegradable. Keep your cat indoors to prevent her from killing birds. Don't walk your dog near waterways (and wherever you walk him, pick up his waste). Don't overfeed your pet. Like humans, pets are suffering from an obesity epidemic. Typically we now feed them two or three times the amount of protein they need, and it is the protein production that is most damaging to the planet (by feeding them less protein and a higher-quality food, they will produce less waste). Ask yourself if you really need all those pet toys, clothes and other paraphernalia adding to the tidal wave of carbon emissions and overfilled landfills. Please."

"'This is really about a commercial system we have created," notes Patricia Cameron, executive director of Green Calgary, a non-profit making the effort to reduce greenhouse gases. "When you look at dogs, their needs are very, very simple."'

And there you have it. Allow your pet to lead you to a simpler, more climate-conscious lifestyle. It will do you both, and the planet, a world of good!

Reflection:

How might you reduce the carbon footprint of your pet or a loved one's companion animal?

Example from Margaret: Make or buy biodegradable toys made from natural fibers.

Research: Vegan Diet Crucial to Reduce Climate Change

As I've said before, personal and planetary wellness are one and the same.[150] Recent research[151] by Oxford scientists published in the Proceedings of the National Academy of Sciences (PNAS) indicates:

"Transitioning toward more plant-based diets that are in line with standard dietary guidelines could reduce global mortality by 6–10% and food-related greenhouse gas emissions by 29–70% compared with a reference scenario in 2050."

The study has some caveats, including details of how we might measure the environmental and economic benefits, but hey, a 29-70% reduction in food-based emissions is not to be trifled with. We cannot afford to ignore the science any longer. The study includes a breakdown of types of diets and their emissions reduction impact:[152]

"When it comes to climate change, following dietary recommendations would cut food-related emissions by 29 percent, adopting vegetarian diets would cut them by 63 percent and vegan diets by 70 percent."

There are burgeoning ill effects of animal agriculture,[153] not the least of which are methane pollution[154] and the epidemic proportions of obesity and heart disease.[155] On the plus side, more and more Americans are choosing to be vegetarian and vegan.[156] Driven by consumer trends[157] that favor vegan eating, investment in the plant-based food industry is increasing. Plant-based food companies have recently set up their own trade association,[158] venture capital fund,[159] and a new nonprofit focused on sustainable foods.[160] According to One Green Planet news,[161] New Crop Capital, the new venture capital fund, has

"…announced a slew of investments in vegan, plant-based and culture-based food companies like Memphis Meats,[162] Lighter,[163] The Purple Carrot,[164] Lyrical Foods,[165] Gelzen,[166] Beyond Meat,[167] Sunfed Foods,[168] and Miyoko's Kitchen.[169] New Crop is in good company, because the opportunity evident in this emerging food space has also caught the attention of many of today's most forward-thinking investors. Firms like Google Ventures,[170] Khosla Ventures,[171] Bill Gates' Gates Ventures,[172] SOSventures,[173] and Horizons [Investments][174] are investing millions in companies that are creating innovative products that make animal agriculture obsolete."

It's time to join the most effective movement for planetary and personal wellness: changing your diet! Vegan recipes are blossoming all over the interwebs, including on One Green Planet,[175] VeganEasy,[176] The Colorful Kitchen,[177] and NutritionCity.[178]

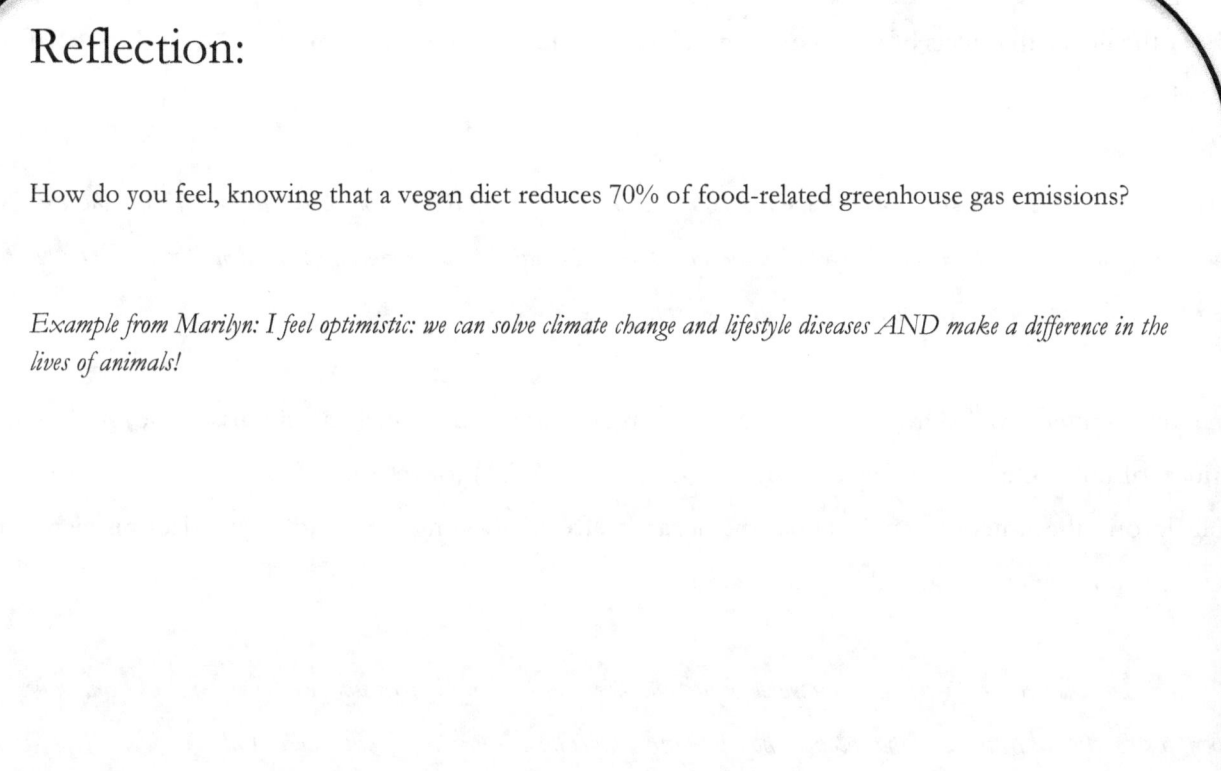

Reflection:

How do you feel, knowing that a vegan diet reduces 70% of food-related greenhouse gas emissions?

Example from Marilyn: I feel optimistic: we can solve climate change and lifestyle diseases AND make a difference in the lives of animals!

Stanford M.D. Endorses Plant-Based Diet

I've written previously about the benefits of a plant-based diet for productivity,[179] climate change,[180] learning,[181] and wellness.[182] Recently, Stanford Professor of Medicine Randall Stafford[183] spoke up about the ill health effects of eating meat via a letter to the Journal of the American Medical Association (JAMA):

"The health benefits of specific components of plants have been documented, as have the harms associated with constituents largely unique to meat," he wrote. "Vegetarian diets have been associated with a reduction in cardiovascular disease mortality by as much as 29 percent and cancer incidence by 18 percent."

In a recent interview,[184] Dr. Stafford, who is an expert in preventive medicine, discussed the letter, which is critical of the United States Department of Agriculture (USDA) for releasing dietary guidelines that do not fully inform the American public about the negative effects of eating meat. He summarizes his views on the relationship between meat and human health:

"People who consume meat generally have worse health outcomes, particularly in terms of heart disease, stroke and cancer. On the flip side, clinical trials show that people who eat mostly plants have better health outcomes. And the evidence goes further than just suggesting an association — it shows that plant-based diets directly cause better health."

Dr. Stafford makes a further point in the same interview[185] about dietary guidelines, (emphasis added):

"It would be much more direct to simply tell consumers to eat less meat. And that would be the most effective way to reduce the consumption of saturated fats.

Despite the tendency of consumers to be attracted to fad diets, I think Americans are more ready than ever to hear a simple recommendation to eat less meat. The dietary evidence is stronger today than it's ever been. And I think consumers are also uncomfortable with both the environmental impact of their diets and the issues surrounding the ethical treatment of animals. The time is right for the USDA to be more direct in their recommendations, even if it means making a recommendation that is contrary to the interests of some entrenched food manufacturers."

The elephants in the room here are the powerful meat and dairy lobbies,[186] which tend to influence USDA dietary guidelines due to their vested interest in keeping people addicted to meat and dairy products. Dr. Stafford suggests[187] some strategies for scientists and consumers:

"I certainly think more pressure from scientists to have the USDA state the obvious consequences of the data would help. I also think it's important that consumers complain to the USDA that the guidance is not nearly as clear as it could have been."

At Alchemus Prime, we advocate a more direct approach,[188] nurturing integrated solutions at the individual, team, and community levels. We encourage and help clients to implement immediate behaviors that have multiple benefits; eating a plant-based diet has positive impacts for climate change, animal welfare, water scarcity, and land degradation. Changing your own dietary habits over time, specifically shifting away from meat and dairy[189] is your way of doing the right thing for yourself and all life.

Every meal is an opportunity for (r)evolution.

Reflection:

Who in your social circle might benefit from this medical opinion? Share this article with them and write their response here.

Example from Marilyn: I shared this information with Margaret, because she is a medical doctor. She responded by saying that this was a very positive sign: doctors are realizing and speaking openly about the importance of a vegan diet.

Meat Substitutes' Carbon Footprint

For meat-lovers who are looking for ways to eat healthier, be kinder to animals, and reduce their carbon footprints, plant-based meat substitutes[190] can be a helpful and convenient option. Recent research[191] by Maximo Mejia of Andrews University[192] suggests that meat-eaters can continue to enjoy the flavors they like while reducing their greenhouse gas footprint by up to ten times:[193]

"While the pollution generated to produce a typical 8-ounce steak is equivalent to driving a small car for about 29 miles, replacing that steak with the same weight of a vegetarian meat substitute is like driving the same car just three miles. Across the board, meatless alternatives are associated with substantially lower emissions than actual meat, according to an analysis of the environmental impacts of 39 meat substitutes presented at the American Society for Nutrition Annual Meeting during Experimental Biology 2016."

Of course, this is good news for meat-eaters, but also for plant-eaters. However, Alchemus Prime's win-win approach[194] asks us to go beyond only carbon footprint and look at other factors for personal and planetary wellness:[195] are these substitutes non-GMO, organic, gluten-free (find out why),[196] and minimally processed?

Often, the answer is no. As a vegan, I occasionally experiment with meat substitutes to see what tastes good. But mostly, I prefer organic foods in their natural form like fruits, vegetables, legumes, grains, beans, and nuts that aren't processed too much. I relax these preferences sometimes when eating out, and find that I'm spoiled in California with a plethora of vegan and gluten-free options. This may not be true for different parts of the U.S. and the world, so simple, whole-foods options might be best.

As we move toward creating a more compassionate, sensible, healthy, safe, and just future, it's important to consider the greenhouse gas emissions and cruelty on our plates. There are also toxins, hormones, antibiotics, bacteria, and so many other undesirables to think about when eating animal products; a simple way to choose wisely might be to add more colorful, organic and fresh fruits and vegetables to your plate. Eat plant-based foods for a safer climate,[197] better health,[198] improved productivity,[199] liberated animals,[200] and you can save money too.[201]

As common sense prevails, vegan meats are becoming the current largest trend in the tech industry.[202] Here's how one individual describes our current predicament:[203]

"The two big questions are, how do we feed 9 billion people by 2050, and what can we do about climate change?" asks Good Food Institute Executive Director Bruce Friedrich. "Plant-based...products are the answer to both of these questions."

Reflection:

Knowing that a product like tofu takes about 10 times fewer greenhouse gas emissions to produce, how might you include more plant-based meat substitutes into your diet?

Example from Marilyn: I experiment with flavorful recipes for tofu using different marinades and sauces.

Tax on Climate-Damaging Foods

The Danish Council of Ethics has released a report entitled *The Ethical Consumer: Climate-Damaging Foods.*[204] The report details the greenhouse gas impact of the livestock sector: 41% of the 14.5% of greenhouse gas emissions emanating from the livestock sector is made up by beef production, and another 20% by dairy cattle.

The Danish Council of Ethics makes several recommendations[205] in this report, including:

- A tax on beef and other climate-damaging foods, accompanied by a subsidy on the least climate-damaging foods.
- Taxes that help reduce food waste.
- Mandatory meat-free days in public institutions.
- Subsidies for farmers wanting to convert to more climate-friendly production, largely fueled by the new taxes.

In general, the Council supports regulation to reduce the consumption of climate-damaging foods, and leave the choice of what to eat with the consumer. This could result in a shift in consumer choices due to changing prices in favor of climate-friendly foods. The Council stated that giving up beef is "unproblematic."[206]

Even with such a progressive outcome, the Danes continue to have dissent within their leadership, with members who believe that human interference in climate change is not unequivocal. Living dangerously in the Anthropocene era,[207] and in the throes of anthropogenic climate change,[208] or human-caused climate change, I find it surprising that the science hasn't sunk in yet for some of us.

However, the tide is turning, and decisions like this are guiding the steps we take to live more in harmony with our life-support system. Kudos to the Danes for their leadership. We support this solution for consumers, farmers, and Ma Earth.

Reflection:

What do you think it would take for a tax on climate-damaging foods to be approved in the United States?

Example from Marilyn: Widespread and collaborative policy and grassroots efforts that extend beyond political affiliations.

Meat and Climate Change

Exploring the connections between meat and climate change is not a new effort. Research from 2009[209] indicates dietary changes can substantially help address global climate change:

"By using an integrated assessment model, we found a global food transition to less meat, or even a complete switch to plant-based protein food to have a dramatic effect on land use. Up to 2,700 Mha of pasture and 100 Mha of cropland could be abandoned, resulting in a large carbon uptake from regrowing vegetation. Additionally, methane and nitrous oxide emission would be reduced substantially. A global transition to a low meat-diet as recommended for health reasons would reduce the mitigation costs to achieve a 450 ppm CO2-eq. stabilisation target by about 50% in 2050 compared to the reference case. Dietary changes could therefore not only create substantial benefits for human health and global land use, but can also play an important role in future climate change mitigation policies."

More recent research[210] shows that only 6% of those sampled in the U.S. understand the remarkably high greenhouse-gas reducing effects of a low-meat diet. Analysis[211] of this study by one of its authors suggests that while communicating the sizable effect of reducing meat in the diet on climate change mitigation efforts might be helpful in motivating environmentalists and those who already eat less meat, it's less salient for heavy meat-eaters.

As we advocate and practice through our services, current research[212] supports the collective approach, or cultural change. A positive, empowering approach to dietary change has the power to reduce climate change and improve wellness, which is the foundation for why Alchemus Prime, and our Diamond Model,[213] was created. It is heartening to see more research on this topic. I'll sum up with the eloquent words of study co-author Annick de Witt:[214]

"As many authors have argued, the greatest potential for a shift towards sustainable lifestyles is through a change in culture and worldview—a shift in assumptions about human nature, our relationship with the (natural) world around us, and our aspirations for the 'good life'. Food touches on social habits and norms; plays a role in mediating power and status; is often key to social participation and acceptance; and is expressive of collective values and identity. Consumption and lifestyles therefore tend to be shaped more by people collectively than individually."

As we combat climate change together, it makes sense to take the most effective action: changing our diets.[215] A wonderful side effect is that we can simultaneously address epidemic lifestyle diseases such as diabetes, heart disease, and obesity.[216]

Reflection:

How do you think more Americans could understand the connections between meat and climate change?

Example from Marilyn: Encourage mainstream climate leaders like Al Gore and Bill McKibben to talk in detail about this topic in front of their huge audiences.

How to Be Happy AND Fight Climate Change

I read several articles recently and two were about the same solution for two seemingly very different challenges facing society: being happy, and improving climate stability.

The first article[217] is about a monk who scientists at the University of Wisconsin call the happiest man in the world. His gamma wave levels, as measured by these scientists, are the highest of anyone studied scientifically. Gamma waves form a metric for cognition and emotion,[218] and seasoned meditators tend to exhibit high gamma waves and more "consciousness" and attention.

So, the happiest man in the world, Matthieu Ricard, states[219] that his joy comes from aligning his values with his actions. Specifically, he cares about all living beings (value), so he doesn't harm animals through a vegan diet (action).

Now, we know from peer-reviewed science that a vegan diet can reduce climate change and improve health.[220] What's interesting about the second article[221] I want to highlight is that it's from CNN, a mainstream news source. George C. Wang, Ph.D., M.D., an integrative physician with very impressive credentials (can you say Columbia University Medical Center, and Johns Hopkins University School of Medicine?),[222] wrote an opinion piece on the fact that being vegan can save the planet from climate change.

Of course, the perspective is Dr. Wang's, but CNN still invited him to write it.

My point is that a plant-based diet allows everyone to win:[223] planet, climate, animals, us pesky humans and our children, from the smorgasbord of lifestyle diseases that ail us: diabetes, obesity, heart disease, cancer (you get the idea).[224] A healthy diet helps us be healthier and happier, AND save the planet from climate change. Now, that's a smart solution.

I'll leave you with Dr. Wang's crucial closing words:[225]

"It is rare that a single choice of ours can have a broad and decisive impact on the climate crisis. We have a moral imperative to choose and advocate for plant-based diets for the health of our planet and the well being and survival of generations to come."

Reflection:

The single most important action we can take for wellness and climate change is adopting a plant-based diet. List three steps that would make it easier for you, or a loved one, to start making the shift today.

Example from Marilyn: Take stock of kitchen supplies and make a list of vegan alternatives; buy a few; taste test them. Learn how to make vegan recipes. Join vegan Facebook groups and ask for advice on food choices, supplements, and more.

1.

2.

3.

Chapter 3 Notes

_____

[95] Access the full report, entitled "Food System Shock Report," via this article: http://www.news.com.au/technology/environment/bye-bye-birdie-civilisation-will-collapse-in-2040-apparently/news-story/96fed454c53130fc311f2351cb4ce0f9

[96] To learn more about the connections between animal agriculture and food insecurity, read the essay entitled "Going to Hell in a Handbasket or Converging on a Solution?" in Chapter 1.

[97] Robert Goodland and Jeff Anhang's paper, "Livestock and Climate Change" uses a life cycle approach to show that animal agriculture accounts for up to 51% of global greenhouse gas emissions. While this work was disputed, the authors successfully rebutted their skeptics. The paper can be accessed here: https://www.worldwatch.org/files/pdf/Livestock%20and%20Climate%20Change.pdf

[98] Read the article about the importance of behavioral science here: http://www.huffingtonpost.com/anthony-biglan/perhaps-behavioral-science-may-prove-to-be-our-most-important-science_b_6764296.html

[99] See Chapter 1.

[100] Stanford research shows the efficacy of behavioral sciences in obesity prevention: https://stanfordhealthcare.org/medical-conditions/healthy-living/obesity/prevention.html

[101] Behavioral sciences have contributed to successes in sanitation: http://water.stanford.edu/research/water-sanitation-and-health

[102] See our peer-reviewed paper on the successful outcome of a behavioral energy efficiency high school curriculum that also showed health benefits: https://link.springer.com/article/10.1007%2Fs12053-013-9219-5

[103] See my blog on how behavioral energy efficiency programs are going mainstream: http://www.integritusprime.com/behavioral-energy-reduction-goes-mainstream/

[104] Read about the five ways in which behavior change can double world food supply: http://science.howstuffworks.com/environmental/conservation/issues/5-ideas-for-doubling-the-world-s-food-supply.htm#page=1

[105] See the next essay, entitled, "Climate Change: What Would Nature Do?"

[106] See the paper abstract here, indicating the abundant possibilities for regeneration of forests if we switch to a plant-based diet: https://agu.confex.com/agu/fm15/meetingapp.cgi/Paper/67429

[107] See Chapter 1, and Note #105.

[108] See the full article here: http://www.bbc.com/news/science-environment-35039465

[109] Life's Principles are a set of principles for how to learn from nature, using the discipline of biomimicry. Learn more here: https://biomimicry.net/the-buzz/resources/designlens-lifes-principles/

[110] Learn more about how much more efficient it is to produce plant-based foods compared to animal-based foods: https://alumni.stanford.edu/get/page/magazine/article/?article_id=29892

[111] Check out this plant-based medical center: http://www.pcrm.org/barnard-medical-center

[112] Food of Love and our other books are available on our website: http://www.alchemusprime.com/our-books/

[113] Read more about this species, black poisonwood, here: https://en.wikipedia.org/wiki/Metopium_brownei

[114] The plant containing the antidote: https://en.wikipedia.org/wiki/Bursera_simaruba

[115] Read more about the late Steve Schneider on his website: https://stephenschneider.stanford.edu/ and in the essay in Chapter 10 entitled, "Remembering Steve Schneider: Climate Warrior."

[116] Learn more about mutations from the Encyclopedia of Earth: http://www.eoearth.org/view/article/159530/

[117] See the first essay in Chapter 2, entitled "Cancerous Meat: Risk vs. Norm."

[118] Robert Goodland and Jeff Anhang's paper, "Livestock and Climate Change" uses a life cycle approach to show that animal agriculture accounts for up to 51% of global greenhouse gas emissions. While this work was disputed, the authors successfully rebutted their skeptics. The paper can be accessed here: https://www.worldwatch.org/files/pdf/Livestock%20and%20Climate%20Change.pdf

[119] See Note #109.

[120] Animal slaughter statistics are disturbingly high in the United States: http://www.humanesociety.org/news/resources/research/stats_slaughter_totals.html

[121] Check out rainforest loss rate and other details here: http://www.rainforestfoundation.org/commonly-asked-questions-and-facts

[122] See article here on the beef industry's impact on rainforest loss: http://www.onegreenplanet.org/animalsandnature/beef-production-is-killing-the-amazon-rainforest/

[123] See article on the impact of plant-based diets on greenhouse gas emissions: http://www.ncbi.nlm.nih.gov/pmc/articles/PMC4372775/

[124] See article on water footprint: http://iopscience.iop.org/article/10.1088/1748-9326/9/7/074016;jsessionid=18812CC096397FD31CD2063CA4F701ED.c3.iopscience.cld.iop.org

[125] Read about cruelty in the dairy industry: http://www.onegreenplanet.org/animalsandnature/is-there-more-cruelty-in-a-glass-of-milk-or-pound-of-beef/

[126] Heart disease can be reversed with a plant-based, oil-free diet: http://www.health.harvard.edu/heart-health/halt-heart-disease-with-a-plant-based-oil-free-diet-

[127] Obesity and other lifestyle diseases can be reversed with a plant-based diet: http://www.ncbi.nlm.nih.gov/pmc/articles/PMC3662288/

[128] Plant-based diet can reverse type 2 diabetes: http://www.treeoflifefoundation.org/service/reversing-diabetes-2/

[129] Colon cancer risk is lowered on a plant-based diet: https://www.ncbi.nlm.nih.gov/pubmed/1552366

[130] Meat and dairy require more land than plant-based foods: http://www.worldwatch.org/node/549

[131] See full story here: http://www.eater.com/2015/2/16/8048069/un-says-veganism-can-save-the-world-from-destruction

[132] Read the report here: ftp://ftp.fao.org/docrep/fao/010/a0701e/a0701e00.pdf

[133] See Note #105.

[134] This cartoon shows how much biomass farmed livestock and pets constitute in the world compared to humans and wild animals. The result is not what you might think: https://xkcd.com/1338/

[135] Read a Cornell ecologist's take on just how inefficient meat-eating is in terms of resources: http://news.cornell.edu/stories/1997/08/us-could-feed-800-million-people-grain-livestock-eat and also see Note #109.

[136] Extinctions are continuing: http://www.biologicaldiversity.org/programs/biodiversity/elements_of_biodiversity/extinction_crisis/

[137] See Note #108.

[138] Read more about the Alchemus Prime Diamond Model in Chapter 11, and on our website: http://www.alchemusprime.com/model/

[139] We work in ways that address climate change and wellness simultaneously as much as possible: http://www.alchemusprime.com/how-we-work/

[140] The full article on the carbon footprint of a pet: http://www.salon.com/2014/11/20/the_surprisingly_large_carbon_paw_print_of_your_beloved_pet_partner/

[141] Read about the impact of pets in terms of transportation numbers here: http://www.telegraph.co.uk/news/earth/environment/climatechange/6416683/Pet-dogs-as-bad-for-planet-as-driving-4x4s-book-claims.html

[142] See Note #139.

[143] See Note #139.

[144] See Note #139.

[145] Read more about the parasite here: http://www.cdc.gov/parasites/toxoplasmosis/gen_info/faqs.html

[146] Read the full entry on endangered species here: http://science.jrank.org/pages/2467/Endangered-Species.html

[147] Mutual gazing between dog and parent can increase oxytocin levels for both: http://well.blogs.nytimes.com/2015/04/16/the-look-of-love-is-in-the-dogs-eyes/

[148] See our website for more on how we work with the true self, http://www.alchemusprime.com/career-manifestation-program/ and see Chapter 11.

[149] See Note #139.

[150] See Chapter 1 for an overview of how personal and planetary wellness are connected.

[151] Read the abstract and access the full peer-reviewed article here: http://www.pnas.org/content/113/15/4146.abstract

[152] See: http://www.nbcnews.com/health/diet-fitness/vegan-eating-would-slash-cut-food-s-global-warming-emissions-n542886

[153] See Chapter 1, especially the essay entitled, "Going to Hell in a Handbasket or Converging on a Solution?"

[154] Read about how the meat industry affects climate change through polluting greenhouse gases: https://news.vice.com/article/meat-is-murder-on-the-climate-anyway

[155] See Chapter 2 for evidence of how meat is connected to lifestyle diseases.

[156] Increasing numbers of Americans choose to go vegetarian and vegan: https://news.therawfoodworld.com/16-million-people-us-now-vegan-vegetarian/

[157] Consumer trends are in favor of vegan foods: http://www.onegreenplanet.org/environment/consumer-trends-driving-the-rise-of-sustainable-plant-based-foods-in-america/

[158] A new trade association has been set up for companies focused on plant-based products: http://www.prweb.com/releases/2016/03/prweb13250076.htm

[159] A venture capital fund now exists for companies in the plant-based industry: http://www.newcropcapital.com/

[160] A new nonprofit has been formed for sustainable foods: http://www.gfi.org/

[161] See news article here: http://www.onegreenplanet.org/news/new-crop-capital-and-good-food-institute-to-disrupt-animal-agriculture/

[162] Learn more about Memphis Meats: http://www.memphismeats.com/

[163] Lighter is a tool that helps you make better plant-based meals: https://www.lighter.world/

[164] Learn more about the Purple Carrot, a meal delivery service: https://www.purplecarrot.com/

[165] Learn more about high quality dairy-free foods: http://www.kite-hill.com/

[166] Geltor makes gelatin without any animal ingredients: http://gelzen.com

[167] Beyond Meat is a startup focused on vegan foods, and is becoming known for The Beyond Burger: http://beyondmeat.com/

[168] Sunfed Foods makes "chicken free chicken:" https://sunfedfoods.com/

[169] Miyoko's Kitchen makes gourmet plant-based cheeses: http://miyokoskitchen.com/

[170] Google Ventures (https://www.gv.com/portfolio/#life) is backing the vegan company Impossible Foods (http://impossiblefoods.com/)

[171] Impossible Foods and Hampton Creek are two plant-based product-touting companies in the Khosla Ventures portfolio: http://www.khoslaventures.com/portfolio/agriculture-food

[172] The Gates Foundation gives grants for research and development relating to plant-based issues and products: http://www.gatesfoundation.org/

[173] SOSVentures is backing various food startups: https://sosv.com/portfolio/food-x-portfolio/ and

[174] Learn more about Horizon Investments: https://www.horizoninvestments.com/

[175] One Green Planet's vegan recipes: http://www.onegreenplanet.org/channel/vegan-recipe/

[176] VeganEasy provides simple vegan recipes: https://www.veganeasy.org/food/recipes/

[177] The Colorful Kitchen offers a range of vegan recipes: http://thecolorfulkitchen.com/

[178] NutritionCity has vegan recipes with exceptional photography: https://www.nutritionicity.com/recipes/

[179] See Chapter 2, the essay entitled, "Plant-Based Diet Improves Productivity."

[180] See previous essay for evidence of the importance of a vegan diet for climate change reduction.

[181] Learn how plant-based diets can improve children's learning: http://www.healthyschoolfood.org/nutrition101.htm

[182] See Chapter 2, the essay entitled, "Meaty Meals: Get the Beef."

[183] Read more about Professor Stafford: https://stanfordhealthcare.org/doctors/s/randall-stafford.html

[184] Access the full interview here: http://med.stanford.edu/news/all-news/2016/07/5-questions-randall-stafford-advocates-a-plant-based-diet.html

[185] Ibid.

[186] Learn more about meat and dairy lobbies: https://www.theverge.com/2016/1/7/10726606/2015-us-

dietary-guidelines-meat-and-soda-lobbying-power

[187] See Note #183.

[188] See Note #138.

[189] See this set of tips for shifting your diet one step at a time: https://www.downtoearth.org/health/nutrition/adopting-plant-based-diet-one-step-time

[190] Wikipedia offers a list of meat substitutes: https://en.wikipedia.org/wiki/List_of_meat_substitutes

[191] Research that quantifies the benefits of not eating meat: http://www.newswise.com/articles/quantifying-the-environmental-benefits-of-skipping-the-meat

[192] Associate Professor Mejia's contact information: https://www.andrews.edu/shp/publichealth/faculty/maximino-alfredo-mejia.html

[193] See Note #190.

[194] See our website for our win-win framework: http://www.alchemusprime.com/win-win/

[195] See Chapter 1 for an overview of personal and planetary wellness.

[196] Gluten causes myriad problems besides celiac disease, such as brain disorders, gut ailments, addictions, and immune system issues. Read more about why gluten is unsafe: https://authoritynutrition.com/6-shocking-reasons-why-gluten-is-bad/

[197] See other essays in this chapter.

[198] Read about Director James Cameron and wife Suzy Cameron's vegan journey to better health and co-founding of a new vegan school: http://www.integritusprime.com/vegan-health-suzy-camerons-journey/

[199] See Note #178.

[200] For a deeper look at what animals face in factory farms and other human-controlled settings, watch the documentary Earthlings: https://www.youtube.com/watch?v=MwPNpy6TJf8, or pick up a copy of my book, One Friend: https://www.amazon.com/One-Friend-Collection-Marilyn-Cornelius/dp/1523989327

[201] For affordable vegan meal options and recipes, see this list: http://www.chooseveg.com/vegan-on-a-budget-17-easy-affordable-recipes-2

[202] Vegan meat options are leading the charge in the tech industry: http://www.organicauthority.com/vegan-meat-is-now-the-biggest-trend-in-the-tech-industry/

[203] Ibid.

[204] See the full report here: http://www.etiskraad.dk/~/media/Etisk-Raad/en/Publications/Climate-damaging-foods-2016.pdf?la=da

[205] Ibid.

[206] It's interesting that the Council considers giving up beef "unproblematic," when many in the U.S. think that giving up any favorite food is very difficult. Read more: http://www.independent.co.uk/news/world/europe/denmark-ethics-council-calls-for-tax-on-red-meat-to-fight-ethical-problem-of-climate-change-a7003061.html

[207] The Anthropocene is a term for the era that marks human industrial impact on the planet. Read more: https://en.wikipedia.org/wiki/Anthropocene

[208] The evidence for anthropogenic (human-caused) climate change is substantial: https://www.ipcc.ch/publications_and_data/ar4/wg1/en/ch9s9-7.html

[209] We know from scientific research that a low-meat or even better, plant-based diet reduces land use dramatically, and can play a large role in climate change reduction: https://link.springer.com/article/10.1007%2Fs10584-008-9534-6

[210] Read the peer-reviewed paper here: https://www.ncbi.nlm.nih.gov/pubmed/26673412

[211] Read the full analysis here: https://blogs.scientificamerican.com/guest-blog/people-still-don-t-get-the-link-between-meat-consumption-and-climate-change/

[212] Ibid.

[213] See Chapter 11 and our website: http://www.alchemusprime.com/model/

[214] See Note #210.

[215] For a medical perspective on why we should embrace a vegan diet, see essay entitled "Stanford M.D. Endorses Plant-Based Diet," also in this chapter.

[216] Ibid.

[217] Read the full article here: https://www.riseofthevegan.com/blog/veganism-is-the-key-to-happiness

[218] Read more about gamma waves and how they relate to how we think and feel: http://www.pnas.org/content/101/46/16369.full

[219] See Note #216.

[220] See this article: http://www.pnas.org/content/113/15/4146.full and read the rest of this chapter for the climate benefits of plant-based eating, and see Chapter 2 for more evidence on the health benefits of vegan diets.

[221] See full article here: http://www.cnn.com/2017/04/08/opinions/go-vegan-save-the-planet-wang/

[222] Check Dr. Wang's credentials here: https://www.researchgate.net/profile/George_Wang7

[223] See Chapter 1 for an overview of personal and planetary wellness.

[224] For a breakdown of how diet influences lifestyle diseases, see Chapter 2, especially the essay entitled: "Lifestyle Diseases: Prevention is the Cure."

[225] See Note #221.

Chapter 4: Educating For Change

What is Education Really For?

Education is about what we are *doing*, not only what we are thinking or what we know.

I contend that education is really for the following purposes:

1. To produce individuals that practice behaviors which promote personal and planetary wellness[226] as a baseline;

2. To allow freedom for students to learn what they want using their own unique learning styles so they can understand *who they are* and gain a sense of purpose;

3. To allow students to discover and hone their talents and skills in real-world contexts so that they can be of service to the world; and

4. To produce responsible and creative citizens that can invent virtuous and environmentally sustainable systems for all life to thrive.

To achieve these goals, we need a behavioral approach, because it is the current collective behavior of humans (e.g. deforestation, factory farming, driving, and mining) that is harming the planet. Additionally, research[227] suggests that traditional educational approaches, such as increasing knowledge or awareness about an issue, are weakly correlated with behavior change. Fortunately, this is not an obstacle, because behavioral approaches can be infused into the education system with positive outcomes; one example successfully promoted climate change mitigation and health promotion behaviors using a behavior change curriculum for high school.[228]

A better approach than fitting into the current system, according to the famous Buckminster Fuller quote,[229] is to throw a better party, making the existing system obsolete.

In this respect, I agree with David Orr that all education is environmental education.[230] Operationally, this means that students, from the moment they begin school, need to learn about the earth from the integrated perspectives of physics and ecology, energy efficiency and renewable energy, carrying capacity and sustainable agriculture, as well as biomimicry.[231]

Fundamentally, K-12 schools must adopt an approach that focuses on grooming young leaders who understand how their world works, what it needs and how to meet those needs. Much more work is needed at the K-12 level, to get us to this ideal.

The better party is one that creates conditions conducive to life. Around the world, it looks like we're heading for that better party, at least in higher education.

Place-based, project-based, indigenous, experiential, and other learning-by-doing, process-focused approaches are becoming more popular. Higher education in particular is growing more and more enamored with these behavioral approaches, as evidenced by findings from interviews[232] with design thinking students at Stanford University, which seem to match the way TeamLabs[233] in Spain functions, with a focus on real-world business performance instead of grades. The rise of innovative institutions of higher learning like the new IKIAM Amazon University in Ecuador,[234] which aims to create an entire undergraduate and graduate education system founded on learning-by-doing, is also evidence of this shift away from knowledge-based approaches to a behavioral focus in higher education.

Of course, indigenous knowledge systems are the holy grail of learning-by-doing, and we will do well to preserve and harness this wisdom to transform educational systems at all levels.
Ultimately, we must produce creative, healthy, and environmentally responsible citizens who can work together to lead our planet out of crisis by putting humanity back into alignment with the rest of life.

That's what education is *really* for.

Reflection:

Describe an experience in any school you attended that taught you about nature through experience. What did you value most about it?

Example from Marilyn: In my undergraduate capstone class, we learned team building in the forest, building bridges over creeks, climbing trees, and learning to make things out of rope to help each other overcome obstacles. I valued how much this experience humbled me and taught me the importance of teamwork.

Reinventing Academia

I'm a lifelong learner and student, and have spent about 24 years in school. My stint at Stanford University consisted of seven years. The first two years I worked as assistant to the late and inimitable Steve Schneider,[235] who consequently became my doctoral advisor. When he died unexpectedly in 2010, life changed for me and I began to see, without his caring guidance and support, that life in academia was, well, madness.

I was surrounded by people perpetuating a system that hurt them. Here's a laundry list of what I observed about academic culture:

- Academics in research universities focus on several types of work, including teaching, research, mentoring, and serving on committees, but they are ultimately evaluated with the most weight given to the number of publications, so they are always under pressure to publish, and tend to think in terms of publications.

- Scholars, including grad students and professors, don't sleep much and by that I mean about four hours per night.

- Very rarely do academics show emotion, in fact it seems to be a taboo.

- Underneath the stoicism, most academics are overcommitted and working frenetically to keep up with all their projects; at Stanford we call this The Duck Syndrome[236] – it is associated with suicide…

- Academics don't know how to say no and are often overextended.

- Academics tend to be perfectionists who are afraid of reviewers and others finding mistakes in their work.

- As much as scholars may collaborate, there is a hierarchy and set of protocols that can disadvantage students and their creative spirit.

- Grad students are at the bottom of the totem pole; postdocs are in a sort of limbo twilight zone; both constitute cheap and sometimes free labor.

- Being on tenure track is a bit like chemotherapy: it almost kills you so you can have your own life back.

- The pay for postdocs is dismal and discouraging, and even for professors the salary can be meager depending on the field, especially when compared to the long hours they put in.

- Many people in academia are stressed and chronically ill.

There isn't enough explicit discourse about much of the above.

During my graduate student career, I became aware of these issues and that resulted in my departure from academia immediately after obtaining my doctorate. I loved many things about being in a university, and I still do some of them, like teach, mentor, and conduct research, but on my own terms.

I didn't want to participate in a system that made people sick and treated them according to what the hierarchical system dictated. Worse, I didn't want to work with people who didn't love themselves enough to challenge that system. I created a system that would be different. A system that I could thrive in and from which I could support others. That system led to the development of Alchemus Prime.[237]

My perspective was affirmed in an Ozy article[238] about the stress climate scientists encounter and the need for wellness and mindfulness in the lives of scientists. Steve's spouse, Terry, and I were both interviewed for the article.

I would be remiss though if I didn't offer what I perceive to be potential solutions to these challenges, for the sake of not only academics who are suffering, but the institutions too, which I consider to be important hives of learning, ripe for transformation.

Here's what might help, based on the Alchemus Prime Diamond Model,[239] which integrates behavior change, design thinking, biomimicry, and meditation:

- A non-hierarchical platform where collaboration and creativity are valued more than rank and publication.
- Fair compensation that enables academics to make a good living without inordinate amounts of stress.
- Metrics that don't prioritize peer-reviewed publications above all, and that don't necessitate superhuman levels of work, but include unstructured time for creativity, invention, rest, idleness, and play.
- Incentives that incorporate personal wellness and teamwork with professional advancement for students, professors, postdocs, staff…everyone.
- Open forums where suggestions can be made for how to improve work-life balance.

- Workshops for how to balance work and life that include exercises that build self-efficacy, teach biomimetic skills, and integrate design thinking processes.

- Various types of meditation classes and sessions available throughout the day and evening.

- Outdoor seating options for class sessions as well as short outdoor contemplation breaks built into class designs.

- Open fora where people at all levels can discuss their emotions and fears instead of suppressing them.

It's very uplifting to note that some small steps are visible. I have friends at Stanford who are embracing yoga and other practices for their personal wellness, and are sharing those practices with others. The establishment of the Windhover Center,[240] where academics can go to meditate, is a beautiful milestone at Stanford. The Health Improvement Program (HIP)[241] includes some wonderful offerings, and is where I began my love affair with Reiki, but is currently only available to employees…why not scholars?

It's also heartening to know that Dr. Emma Seppälä[242] works on campus and is working within the Stanford system to change it – I've been blogging about her science-based approach to success through happiness,[243] as well as the importance of mindfulness[244] for leadership. Academia is a world of possibility, and we will innovate and achieve at our highest potential when we empower our scholars with balance. Stanford is a hub of impressive innovation; just imagine how that innovation could be enhanced if everyone was healthy and happy too!

Reflection:

Describe a collaborative experience you had while in school or university. What was best about it?

Example from Marilyn: I was part of a team at Stanford that made a video about using less energy. The project took many hours and was very rewarding when our class voted our work best in class, and many members of the public enjoyed our funny video. The best part about the experience was working with very different people, overcoming obstacles like time and personality clashes, and getting rid of hierarchy to work as a fluid and effective team.

Operationalizing Science: A Wider Audience

Building on my essay about how to reinvent academia,[245] I want to focus a bit on the constraints that make scientific research limited in its usefulness. There is so much innovation coming out of academia, but the best of academic tend to bury it with their publishing habits.

For instance, consider the following numbers:[246]

"Up to 1.5 million peer-reviewed articles are published annually. However, many are ignored even within scientific communities – 82 per cent of articles published in humanities are not even cited once. No one ever refers to 32 per cent of the peer-reviewed articles in the social and 27 per cent in the natural sciences."

According to Savo Heleta, at Nelson Mandela Metropolitan University,[247] there are three reasons for this unfortunate trend:

1. Academics may feel it's not part of their role to write for the public, and that this would somehow be beyond or against their intellectual mission. Crucially, academic research must be communicated clearly so it can be applied in real world settings. Fortunately, some wise professionals, like Thomas Hayden at Stanford,[248] have dedicated their careers to science communication. This trend of preaching to the choir, then, is changing.
2. There are no incentives for academics to share their research beyond peer-reviewed realms. This is an unfortunately true tendency, and binds professors, postdoctoral scholars and students alike to a hamster wheel of frenetic publication production.
3. Academics may lack skills for writing for a lay audience. As Steve Schneider, my late mentor, used to teach, academics tend to bury their lead.[249] Training programs, such as Professor Hayden's, are critical in empowering scientists to communicate effectively in the media's biased arenas.

Knowing these constraints, it is important to provide the appropriate moral framework, identity framing, skills, and incentives to encourage academics to share their work beyond their typical peer-reviewed journals, so that society can take action to safeguard humanity in a rapidly changing climate, with human health at peril due to epidemic lifestyle diseases.

Reflection:

What types of scientific articles would you like to see in more accessible language?

Example from Margaret: I want to see research on food and illness in more accessible formats so that communities, especially disadvantaged groups, can understand the implications clearly and take appropriate action quickly to avoid ill health.

Chapter 4 Notes

---

226 See Chapter 1 for an overview of personal and planetary wellness, and how they are inextricably connected.

227 Read more about educational interventions: http://www.tandfonline.com/doi/abs/10.1080/00958969909598627#.VKnEjivF-So

228 Learn more about this behavioral curriculum: http://link.springer.com/article/10.1007%2Fs12053-013-9219-5

229 Read the quote here: http://www.goodreads.com/quotes/13119-you-never-change-things-by-fighting-the-existing-reality-to

230 Read the excellent paper here: http://www.eeob.iastate.edu/classes/EEOB-590A/marshcourse/V.5/V.5a%20What%20Is%20Education%20For.htm

231 See the Glossary and this site for a definition of biomimicry: http://biomimicry.org/what-is-biomimicry/

232 Read the interview highlights here: https://www.linkedin.com/pulse/how-can-we-fundamentally-change-higher-ed-tim-brown

233 Learn more about TeamLabs: http://www.teamlabs.es/

234 Learn more about IKIAM here: http://www.ikiam.edu.ec/index.php/en/

235 Read more about Steve Schneider: http://stephenschneider.stanford.edu/

236 Read more about the Duck Syndrome: http://www.nytimes.com/2015/08/02/education/edlife/stress-social-media-and-suicide-on-campus.html

237 See Chapter 11 for more on how Alchemus Prime was formed.

238 Read the Ozy article about how stressed climate scientists can become: http://www.ozy.com/fast-forward/its-the-end-of-the-world-how-do-you-feel/62757

239 See Chapter 11 for more on the Alchemus Prime Diamond Model, and our website: http://www.alchemusprime.com/model/

240 Read more about the Windhover Center: https://windhover.stanford.edu/

241 Learn more about the Health Improvement Program at Stanford: http://med.stanford.edu/hip.html

242 Read more about Dr. Seppälä: http://www.emmaseppala.com/

243 See essay in Chapter 11 entitled, "Conscious Leadership = Happiness"

244 See Chapter 6 for essays on mindfulness.

245 See the first essay in this chapter.

[246] See the full article here: http://www.straitstimes.com/opinion/prof-no-one-is-reading-you

[247] Read the full article here: https://theconversation.com/academics-can-change-the-world-if-they-stop-talking-only-to-their-peers-55713?utm_source=twitter&utm_medium=twitterbutton

[248] Learn more about Tom Hayden here: https://comm.stanford.edu/faculty-hayden/

[249] Learn how scientists can bury their lead, which means they mention the most important point somewhere deep in an article or not till the very end, and in a way that seems very conservative: https://www.markey.senate.gov/GlobalWarming/files/HRG/052010SciencePolicy/schneider.pdf

Chapter 5: Healing Social Injustices

From Cruelty to Compassion

One of the ways I understand and practice ethics (Kantian ethics,[250] for example) is that when we hold a principle, we make no exceptions, because exceptions violate the moral principle. For example, if we hold the principle of "do not lie," then we do not commit or allow any form of deceit, cheating or lying. If we make an exception, we violate the honesty principle. Of course, many lies have been told for arguably "good" reasons, and anyone who has watched "The Good Lie,"[251] or surprised a loved one on their birthday can appreciate this point.

Which set of moral principles do we abide by? Which exceptions do we allow? How do we decide? Are we deciding consciously?

I recall a conversation with Alchemus Prime Advisory Board member Sailesh Rao, head of Climate Healers,[252] in which he made the case that the fight against animal cruelty is intimately connected to the fight against sexism, racism, discrimination against LGBT folks, slavery, and even genocide. In all cases, fundamentally, some beings are considered inferior to others, and they suffer adverse consequences. We went on to agree that from this moral perspective, a perspective of justice, we could not condone cruelty or injustice some of the time; we were bound by morality to unhook ourselves from the entire system of cruelty. In our case, this means that we are both strict vegans for the climate, animals, children and future generations, our own health, and all life on the planet.

We were talking about why veganism, often called extreme, is actually not extreme at all, unless we think of it as extreme compassion, and extreme justice. From a scientific perspective, given that animal agriculture accounts for as much as 51%[253] of global greenhouse gases, veganism is what common sense would suggest. In a world suffering disproportionately from aggression and inequality, compassion and justice seem to be what the moral doctor would order.

Do we want to take care of our health only some of the time? Does it make sense to support a stable climate only some of the time? What about supporting marriage equality only on Sundays? As Sailesh pointed out to me in that conversation, such an approach seems ridiculous at best.

Others, like my esteemed colleague Barry Bruce, have suggested to me that my strict vegan stance seems somewhat arbitrary, because mushrooms, like mycelia,[254] and plants[255] are conscious and have active communicative networks. To this, the only response I could muster was: does a carrot or mushroom suffer in the same way as a cow or pig or chicken or fish? A friend and dedicated animal rights activist brought this subject up at the San Francisco Cowspiracy[256] premiere after-party; she asked if it was all right to eat plants, knowing that they are harmed. Such questions persist.

I do not have all the answers, but the path I aim to follow is one of sacred connection: to cultivate reverence for the carrot and mushrooms that I eat to live. To take in their energy and nutrients with appreciation, acknowledging that they become a part of my body and help me achieve and maintain wellbeing. As a vegan, I show respect for the living beings I do not need to eat by letting them be. I draw inspiration from the ways that indigenous peoples maintain a sacred connection to all life. In this worldview, there is no separation between humans and the rest of nature. Only what is needed is taken. Violence, if needed, is committed with respect and ritual.

We have strayed far from this way of being. Violence, in particular, is rampant in our everyday lives. I flinch when someone utters a swear word – to me it represents a violent spike of energy, a harmful intent. I jump when someone tugs on his or her dog's leash. Thoughts race through my mind: what if the dog could be free to sniff anything he wants? What if he could be completely free, as I aspire to be in my life? Well, then I would never own a dog again, or live in a place where he could always be off-leash. I cry when a bird or a squirrel or a raccoon is hit by a car or shot.

I want us to build systems that consider and account for the wellbeing of *all* life. I want us to live *with* other animals safely, not in domination of them. I want us to create a reality that is less violent, more loving, and more in tune with our own inner wiring for empathy.[257]

Perhaps there are no absolute answers. However, I believe that we humans have the capacity to imagine and create a more compassionate and more just world for all beings. How we choose to embrace that role, and exercise our collective responsibility, is the question of our time. Let's ask not how we can dominate, control, or benefit from the complex and beautiful fractal of nature, but how we can humbly fit into its sacredness.

I have by no means arrived; I still contemplate these questions and many others. Where do you stand on the spectrum between cruelty and compassion?

Reflection:

What is an action you take consistently all the time, and one that you do not? Why?

Example from Marilyn: I eat vegan all the time, because I have all the options available to me. I don't stay out of cars all the time, because I don't always have a public transit or biking option, and need to carpool or use a rideshare or other service.

Beyond Racism: Creating Identity in a New Paradigm

It was July 11, 2014. As I sat in Cline Library Assembly Hall at the Northern Arizona University listening to Dr. Sandra Fox speak about racism in American Indian education, I realized: it's my story. The details are different, certainly, but the thread of racism is the same.

I was attending the Fifth American Indian/Indigenous Teacher Education Conference in Flagstaff, Arizona. I felt humbled to learn there, as I did at the World Indigenous People's Conference on Education (WiPC:E)[258] last May in Honolulu, about the profound wisdom embedded in indigenous ways of knowing and doing – wisdom that all humans could benefit from as we face an increasingly warming, dis-eased, and perilous world. I feel sad that indigenous wisdom has, over time, been quashed, ridiculed, and forced into a cookie-cutter model of "education."

As a high school student, I felt the sting of racism when my application for a government scholarship was declined. My indigenous friends all received scholarships and went abroad. How did I know I was being discriminated against?

I had the highest test scores in that year in my country.

I am ethnically Indian, born and raised in Fiji. I remember the day I opened the rejection letter and cried; I felt uncertain, unworthy and small. I did not know why, except, perhaps, because I was not indigenous. I didn't understand.

I was wounded but I refused to do what was done to me. I adopted a global identity, never rejecting my identity as a Fijian but transcending the limits of judgment that tend to surround race, ethnicity, and nationality. I embraced humanity and began to value myself as an indigenous earthling with unique talents. I wanted to serve Ma Earth, and I was going to find a way to develop myself so that I could fulfill my purpose.

I knew I could never stop being Fijian. I had to find ways to cultivate a resilient identity, strong self-esteem, and consistent self-worth. I am by no means there yet; it is a lifetime worth of practice. I am driven by the fact that there is no longer any room for racism and injustice in our development as a species. We must transcend this ignorant madness. We are bigger than this. We are wired to connect with each other.

Fast-forward a year. At nineteen, after mixed experiences at the University of the South Pacific in Fiji, I traveled to the United States, seeking college education and acceptance. I found both in the San Francisco Bay Area through exemplary teachers and diverse friends. More than fifteen years later, with unwavering support from my parents, despite political upheavals in Fiji that led to financial hardships, and interfered with my international student visa and my ability to stay in school, I managed to pull through with degrees in graphic design and environmental resource management. Later, I returned to California to earn a Ph.D. in climate change and behavioral science.

Despite these academic milestones, I felt like a misfit in academia; I found the fear of failure, overcommitted and exhausted culture, perfectionism, and hierarchy stifling. Again, I found myself responding by looking inward to discover who I wanted to be and how I wanted to work and grow.

Following my inner compass, I co-founded a company that empowered leaders around the world to face the challenges posed by climate change, lifestyle diseases, and failing educational systems. I realized then with gratitude that I had begun to live my dream. As I formulate a new company now with a focus on leveraging human ingenuity to synchronize business with nature, I continue to work toward my dreams of a better world.

Importantly, I have forgiven those that discriminated against me. These days I am based in California and visiting Fiji when I can to be with my parents and to serve communities there. I have felt appreciated and accepted in my homeland once again through my work, which is a blessing. On the outside, I am a dual citizen; on the inside, I am a citizen of the world.

Crucially, I have accepted myself, despite the labels that others still use for me when they fail to see me. As I continue to explore my identity, learn new skills, and grow as a human being, I realize that I enjoy learning from the teachings of many cultures and traditions. My spirituality is one of connectedness that allows all life to be.

We are living in a new paradigm even as the old one, built upon the precarious sticks of short-sightedness, greed, violence and injustice, crumbles. This time is one of sharing, compassion, creativity, and above all, collaboration.

The next time you feel devalued, show compassion for that person who isn't able to see you, and remind yourself of your identity based on who you know yourself to be, and how you know you can grow.

No one else can define you. Create yourself based on your passion, purpose, and service to the world!

Reflection:

Think about the last time you encountered racism against you or someone else – how did you feel?

Example from Marilyn: I saw racism happening in a movie trailer and I got a sinister feeling in my belly. Also a dull, sick feeling. I also felt reminded that there's still a lot of work left to do to eradicate racism.

(No) Scientific Basis for Racism

Was it Ferguson that started it? Then Baltimore. Then Charleston…

Well, no. Racism has been around for a very long time.[259] It's just become an explosive media topic recently.

I've written about my own experiences with racism,[260] explored nonviolence in jail, and learned that through the lens of emotion, we are all the same. Racism has become so prevalent that it's time to figure out effective solutions and implement them. First, let's recap.

Racism is about fear masquerading as insecurity, anger, and territoriality. It's about identity. It's about hate. It's about "the other."

Despite the fact that Darren Wilson, the police officer who shot Michael Brown, will not be going to jail because the Justice Department can't prove racism was the motivation, the laundry list of systemic racist acts against African-Americans in Ferguson is downright disgusting.[261]

The Baltimore riots brought to the forefront the long and horrific history of racism there, highlighting the poverty and segregation[262] that stifles black residents, especially youth.

Then we woke up to the shocking murder of nine black people[263] at the hands of 21-year-old (and white) Dylann Roof in Charleston. As if we could handle any more of this collective pain and the historic symbolism that reminds us of white supremacy.[264]

Racist violence, or what's being called 'racial terrorism' in the case of Dylann Roof, isn't the only type of explosion happening in the media. We have become so sensitive to race that every event perceived as a violation of racial identity receives a massive backlash. One classic example is Rachel Dolezal, the white woman who identifies as black and had the whole country in an uproar. When I first heard about her, I realized that for whatever reason, she is choosing to deny her racial identity and adopt another. There must be trauma, I thought, and recent news suggests there was.[265]

What's interesting is not only that Ms. Dolezal identifies as transracial, but also she is attempting to transcend the trauma of her childhood and adopt an identity based on values, not skin color or racial heredity. Some important questions this raises for me include:

- Is identity, in fact, malleable and flexible?
- Can we identify as we choose?
- Who gets to decide if a person can be transracial
- What about being transracial to help a community?

Must we be put into boxes based on stereotypes, as in the film "Black or White" starring Kevin Costner? Watch the movie, read the scathing Forbes review[266] and decide for yourself.

What is to be done about racism? I grew up in a diverse community and attended an international high school where thirty-three nationalities were represented. Diversity was like breathing the air that was everywhere. When I don't see people who are different from me, I feel that something is wrong.

I learned empathy, compassion, tolerance, and love for diversity as a child. I was fortunate. Many of us aren't.

We can look to science for answers. Although "diversity" is used to describe the many cultures of our world, it seems we often concentrate on a range of differences rather than similarities. There is not one trait, gene, or characteristic that distinguishes all members of one race from all members of another race. If you map any number of traits, none would match the accepted idea of race, because modern humans haven't been around long enough to evolve into different subspecies and we've most often moved, mated, and mixed our genes. Beneath the skin, we are one of the most genetically similar of all species. In fact, research shows that, based on DNA analyses, ALL MEN descended from a common ancestor, the so-called "genetic Adam."[267] The only exception is an African-American man from South Carolina called Albert Perry, and his DNA suggests interbreeding between archaic and modern humans.[268] Yes, humans. It's who we all are.

Population geneticist Spencer Wells,[269] a former fellow Stanford scholar who went on to write *The Journey of Man*,[270] now heads the famous Genographic Project,[271] which aims to trace the Y-chromosome and ancestral human migration out of Africa. His work beautifully highlights the fact that our DNA is a thread connecting all people around the world through history.

With this scientific evidence in mind, we can and should look for those golden threads that weave us all together, those attributes that make us similar, such as our physiology, our emotions, and our needs for acceptance, belonging, love, and happiness, rather than those that we have used so often to segregate and divide, such as skin color, religion, and race. Especially now, when the United States is engulfed in a political deluge of alternative facts and explicit racism, it's important to look at the positive scientific foundations for compassionate and accepting behavior.

There are science-based tools that can help us achieve harmony and celebrate diversity. We can turn police officers into "peace officers." Those tools include mindfulness meditation, creative approaches to behavior change using design thinking and biomimicry, and above all, leadership through integrity – finding our own identity and being true to it, without needing to negate or put down how anyone else identifies.[272]

These tools are transformative and they offer win-win solutions, but the seed of change is, simply, love. Can we love ourselves? Can we love our traumas? Can we embrace our imperfections, mistakes, and flaws? Can we reflect on our choices and make more constructive ones? Can we be mindful of our habitual thoughts, and think better ones? I witnessed incarcerated men doing all of this in the San Francisco County Jail,[273] and was floored by their wisdom, presence, and charisma. Change happens when we go to the root of the issue. What do we fear? How might we transform that fear into love?

My business partner made important contributions to this essay.

Reflection:

Aside from our genetics, what other commonalities tie all humans together?

Example from Marilyn: We all bleed, we all hurt, we all love, and we all die.

Seeing Beyond Orange: My Weekend in Jail (Part I of II)

I spent a weekend in the San Francisco County Jail.[274] No, it's not what you think. Wait, what did you think?

I've been interested in the Alternatives to Violent Project (AVP) for about two years now. Thanks to my friend Ben, who is a trained AVP facilitator, I was part of a small team of participants attending a workshop focused on nonviolent communication (NVC) on June 13 and 14, 2015. This post will cover some of my experiences from the first day.

Day 1

We reported to the jail at about 7.45 a.m., which meant a 5.45 a.m. start for me. We were armed with warm layers of clothes, meals, snacks, water, and curiosity. Everything else, including cell phones and other valuables, which qualifies as "contraband," was not allowed. After surrendering our IDs and receiving visitor passes, we walked past the visitor booths (the kind you see in the movies, with tiny cubicles and phones) through the first of a series of doors.

It was my first time inside a jail, period. I had no idea what to expect.

At first, it felt a little surreal. And sterile. The corridors were long, wider than a hospital's, and empty. We walked to the cafeteria to put our food in the guest fridge, and then walked to the "pod" where we were going to spend the next twelve hours. A pod is a section in the jail that houses about fifty inmates. It's designed with a curvature, apparently based on Jeremy Bentham's Panopticon,[275] which allows all cells to be visible to the jail staff.

Walking into the pod was definitely uncomfortable for me. I'd never seen so much orange. I could see into every cell, and it felt like I was invading a privacy that inmates didn't have. The thing I found most disturbing was when an inmate stood in front of the locked glass door to their cell and looked outside. It reminded me of zoos. I looked away. We walked into the "gym," which consisted of a basketball rim in a high-ceiling concrete room with too much ventilation. It was breezy and chilly.

We arranged chairs in a circle, put up posters of NVC guidelines and workshop agreements, and then sat down.

The inmates participating in AVP were released from their cells and began to file in. They wore orange t-shirts, sweaters, pants, socks and shoes, except a few guys who wore white shoes. Our pod consisted of mainly African-American, Latino, and other minority races.

We started off with an introduction game where we had to come up with an adjective and gesture to go with our name. I was Mellow Marilyn (making wave-like motions that some thought was hula) and Ben became Building Ben. Happily, this game was tremendously helpful for remembering all twenty-seven names in the group. And, it broke the ice.

Sitting next to different inmates during the first half of the day, I noticed that they were hyper vigilant. They noticed EVERYTHING. One example that really struck me was when an inmate saw a *mosquito* on my *black* jacket sleeve and came over from about three feet away to gently flick it away. I hadn't even noticed it! I was super grateful because I am allergic to mosquito bites.

After lunch, we did an exercise where we silently wrote down how we react when we are angry and fearful, and taped our answers to the front of our bodies. We then walked around reading each other's responses. When we later discussed this exercise, a realization jumped into my mind: *Silence is Vulnerability's ally.*

We had shared our deepest vulnerabilities with each other nonverbally. Somehow, the silence equalized us and helped us respect one another; we had also begun to trust each other a little.

We followed this exercise with body sculptures. We were each given a statement, first negative, such as "I don't exist" or "I can't succeed" and asked to individually express this, using our body as a sculpture. The resulting sculptures were all stationary, slumped, and sad. Then we flipped the statements into positive ones: "I do exist!" and "I can succeed!" These expressions transcended sculpture and turned into dances, victory laps, fist bumps, and hugs.

The oldest inmate in the group, Lightning Logan* pointed out during our debrief of this exercise that human emotion, and its expression, is universal. We all behaved in similar ways regardless of our race, gender, age or other differences.

A second insight started buzzing in my brain: *When we examine ourselves through the lens of emotion, we see we are more alike than different.*

For one exercise, we were asked to talk about an incident that led to violence. I shared that my very first boyfriend had pulled a knife on me once. I had wanted to end the relationship because I was feeling very restricted by his jealousy and possessiveness. He threatened to kill me, and himself. I was 18. The inmates in my group expressed their surprise at hearing this. We proceeded to talk about the importance of valuing ourselves enough to make better choices. One of them, Alternative Adam,* turned to me, smiled, and said reassuringly, "Not all men are like that."

Towards the end of the day, through our discussions, it became clear that every inmate was here because of decisions he had made (e.g. to get into that gang car, or to hit that person, or to sell those drugs), and in many cases, due to anger that he had not been able to control. What ran much deeper, and what Building Ben and I talked about that evening but were much too tired to discuss fully, was our privilege, versus the inmates' volatile social and physical environment, the unfair society that they were born into, and the way incentives were aligned to encourage quick riches and short cuts. These situations had almost set them up to fail and end up in this system of punishment. *Can we realistically be expected to do something different while violence is the only option we know?*

To me, AVP felt like a lighthouse in the storm of that punishment, illuminating previously unknown and unseen pathways to achieving goals, to shedding old role models and adopting new ones, to creating a future that does not need to be influenced by the past. It's not easy, but for those who are ready for change, we have the power to make it happen. Together, sharing our pain, anger, and fear, and opening ourselves to something better, many new outcomes and futures become possible. I saw that some inmates were beginning to see the light, and that gave me a warmth that I recognized as hope.

By 11pm on Saturday, in my jammies on Building Ben's couch, I was beyond exhausted. Three hours of sleep the night before, and twelve hours of participating in the workshop had taken its toll. I passed out and got six precious hours of deep sleep, waking the next morning feeling like a new person. It was time for Day 2.

*Names (except Ben's and mine) have been altered to respect the privacy of inmates, participants and facilitators but convey the flavor of how the naming exercise helped us bond.

If you are interested in serving in the SF County Jail through AVP, contact Ben Glass by email (ben@talkwise.com) or phone (415-336-4555). For general information about AVP, go to their website.[276]

Reflection:

Recall a time when you experienced violence against you or someone else. How did you react? Why?

Example from Margaret: I was ready to punch a drunk guy who nearly pushed my daughter and I over the sea wall, while we were sitting and enjoying the ocean breeze. I reacted in anger but also with a strong desire to protect my daughter. Luckily, others had also witnessed his behavior and called the police and they arrived shortly.

Seeing Beyond Orange: My Weekend in Jail (Part II of II)

In the previous essay, I wrote about the first day of the Alternatives to Violence Project (AVP) workshop on nonviolent communication (NVC) that I participated in at the San Francisco County Jail. This essay covers my experiences on the second day, Sunday, June 14, 2015.

Day 2

It was about 8.30am. We greeted each other with our adjectives and names. Mellow Marilyn was feeling like she was in a community of friends that morning. That's because she was.
Our activities took us deeper into our chosen themes of fear, anger, low self-esteem, and poor communication skills.

First, we discussed responses to fear, and soon enough I noticed that some of the responses we were suggesting for the given scenario: a 14-year-old female is at home alone and a seemingly unstable male arrives at the door looking for someone to retrieve a loan from, included lies, like "Let me get my Sheriff Uncle from upstairs."

My integrity meter began to flare, because the guidelines for NVC state that we must take a position of truth. I brought it up and we discussed the intersections of fear and authenticity, as well as honesty and safety. I realized that as a human being and as a facilitator, I'd be pondering this issue for a long time to come. *How might we address fear in ways that have integrity? How might we be honest and stay safe in dangerous situations?*

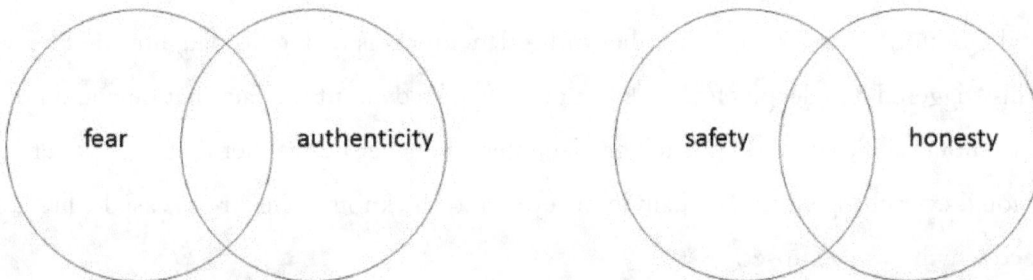

What are in these intersections between fear and authenticity, and between safety and honesty? How might we navigate these territories in nonviolent ways?

After lunch I noticed that one of the inmates, Quintessential Quinton*, who had been on the phone when I went outside to have lunch, was extremely upset. I smiled at him and we communicated non-verbally from across the gym. I sent him supportive glances and after an hour or so he relaxed and began to participate again. I wondered what had happened to him. Later, I would find out.

Sunday was a colder day, and by now our venue, the gym, had become an icebox. Quintessential Quinton* watched us don all our layers, share blankets from the cells, and commence shivering, and then called it. He spoke up, saying it was time to move. Hallelujah!

We moved into the classroom, which is much smaller and intimate. We shed our jackets and continued with our workshop much more comfortably.

In the next phase, we began to share the choices, circumstances, beliefs, interactions, and relationships that had brought us to our current situations. We analyzed emotional triggers and traced them back to the first incident that had embedded the triggers in us. This was powerful, and we helped each other with strategies for how to be non-reactive.

For me, feeling like other people are not respecting my time is a huge trigger, and it goes back to childhood when my dad would be very strict about being on time. I was offered a gift by one of the facilitators, Alright Alexandra.* She reframed the problem for me: "You could use the extra time you spend waiting for someone as if it were a gift, the way an exhausted new mother might." I realized that I could use unexpected free time to meditate, read, or enjoy my surroundings. I was reminded that 90% of what happens to us is how we respond to it. I began to think of ways to apply this wisdom to my current dilemmas, choosing to drop expectations in situations.

Quintessential Quinton* shared that the mother of his daughter was telling his daughter that he was not her father, and this triggered him deeply, fueling his anger. We talked about the fact that he could not control this woman's actions; all he could do was tell his daughter whenever he saw her that he was her father and that no one could ever change that. The pain in his eyes, and the knots in his brow, eased a little, but only a little. I realized why he was so upset.

It started to sink in for me: many of the predicaments inmates face are incredibly complex and painful. Being in jail, indoors with no sunlight, and being expected to change one's mindset and behavior when surrounded by jail politics, gangs, and hierarchies, is very, very challenging.

Later, in the Lifelines exercise, we mapped out our current age, the age we expected to die, our values, goals, and action steps. We then sealed the envelope and dated it for the day we wanted to open it up again and see how far we came.

We realized through our discussions that violence, and the events that could lead to violence, such as arguments, were often distractions from our goals. Once we had set goals, we could constantly evaluate situations with respect to our goals, and make better decisions; constructive decisions that would promote nonviolence. This is what I took away: *Having clear goals gives us the structure we need to help avoid violence and keep our eyes on the prize.*

We wrapped up the workshop with a very powerful exercise called Acknowledgements. In pairs, we first listened to what our partner wanted to be acknowledged for, and from whom, then we took on that role and verbally delivered the acknowledgement. The revelation for me was that I expect a lot from myself, and push myself to do more, be better, and give more. But, I spend very little time acknowledging myself. When my partner, Marvelous Miguel,* acknowledged my progress in self-development and releasing of old patterns, my heart began to expand with gratitude. *How might we spend more time acknowledging and uplifting ourselves and each other?*

One of the inmates, Resilient Richard, broke down in tears after the Acknowledgments exercise, stating that hearing an acknowledgment from his partner, who was role-playing his father, made him realize how much he misses his father. This was a beautiful moment, where all the rest of us, inmates, participants, and facilitators alike, held space for this brave man to cry for his father and his children, and vow to make better choices that would get him out of jail for good, and fulfill his dreams.

After some quick written feedback on what we had learned in the workshop, it was time for graduation. We didn't have music, so some of us hummed cheerfully while certificates were presented. We did victory laps, shook hands and hugged as we celebrated the strong bonds that had been built in just twenty hours of time together. Our last ritual together was a moving circle in which we bowed and said Namaste to one another. The Divine in me sees the Divine in you. YES.

While the facilitator team debriefed in the classroom, Aspirational Amanda,* a fellow participant, and I hung out with a few inmates at a table in the common area of the <u>pod</u>. We chatted about our relationships with our parents and how they had instilled positive and negative patterns in us, and shared our process of taking charge of our lives while continuing to feel tremendous love and gratitude for our loved ones.

This conversation continued as we also started playing Last Card. I rarely play cards but when I do, it's typically with my family. I noticed how relaxed I was feeling, surrounded by inmates in orange who now felt like part of my family. Aspirational Amanda,* Laid Back Larry,* and Resilient Richard* were sitting with me, and Braniac Bobby* looked on.

I won the first game and as we kept playing; the inmates started joking with us. Here are a couple of the jokes:

Resilient Richard* to Aspirational Amanda*: "Now we're going to play for your watch; this IS jail, you know!"
Laughter.

Braniac Bobby* to me: "Is your backpack made of hemp?
Me: Yes.
Resilient Richard*: Can we smoke part of it?
Me: No!
Aspirational Amanda*: Well, if they were to smoke part of it, which part would you give them?
Me, after thinking for a bit: The front pocket."
Braniac Bobby* comes around to study my little backpack's front pocket. We laugh hysterically.

We started talking about our favorite games from childhood, and mine, Scrabble, was brought out. Before we could begin, however, the deputy called us away to stand behind him, killing my excitement. He was concerned for our safety.

That took me back to a whole different kind of surreal. The difference between Saturday morning and Sunday evening was world-sized. I couldn't believe I was leaving, and not returning tomorrow…

I left the jail feeling connected to humanity, healing in ways I could not have expected, grateful, humbled, and deeply moved. We had been our real selves with one another, and that was priceless.

I feel so grateful for this opportunity to learn from men who have faced situations that I don't even have a frame of reference for. Thank you, guys, for being strong. You will prevail.

—

*Names (except mine and Ben's) have been altered to respect the privacy of inmates, participants and facilitators but convey the flavor of how the naming exercise helped us bond.

If you are interested in serving in the SF County Jail through AVP, contact Ben Glass by email (ben@talkwise.com) or phone (415-336-4555). For general information about AVP, go to their website.[277]

Reflection:

What would you like to be acknowledged for? Write a note to yourself acknowledging these qualities.

Example from Marilyn: I am in appreciation of my ability to stay true to myself even when it's really difficult and my friends, family, or colleagues don't agree with my stance.

Racism and Cecil: Making the Connection

Not too long ago, the Internet was ablaze with blogs, memes, and mass outrage about Dylann Roof and the Charleston Church Shooting.[278]

Then it was Cecil the lion, hunted and killed "for sport" by dentist Walter J. Palmer, and the New York Times called the outrage "Internet Vigilantism."[279]

In the animal rights and vegan communities, the backlash is a little different, as exemplified by memes depicting Cecil next to a factory farmed animal, and comments about the hypocrisy of defending one animals when we eat so many.

Climate Progress released a blog post[280] about the science of why we are all so upset about Cecil – essentially because we tend to like charismatic megafauna (popular large animals).

People have started to connect these last two incidents of mass outrage by asking why we care so much about Cecil and so little about black people.[281] However, I contend that this is not the question we should be asking, because false dichotomies create animosity when what we need is common ground. The question is: can we connect the dots here?

The Charleston shooting, Cecil's murder, the hate that the LGBT movement has faced in the past, the torture of animals in factory farms, the current blatant political stance against Muslims and Mexicans, and many other such atrocities are linked through one issue, summed up aptly by this quote from doctor and humanitarian, Paul Farmer:[282]

"The idea that some lives matter less is the root of all that is wrong with the world."

We must make the connection,[283] as the recent book Circles of Compassion[284] does, between the movements for human rights, black lives, women's rights, LGBTQ rights, animal rights, and environmental protection – they emanate from the same source: a need to matter just as much as any other. A need for justice. A need to be well.

Once we learn that these movements are connected, we can unite to achieve a peaceful and loving future filled with community, collaboration, and compassion. It may sound utopic but such a future is possible with the right motivation and science-based tools;[285] the LGBTQ movement has shown us what a unified movement and its victory can look like.[286] It's time to extend that victory not only to all humans, but also to all species. Because all life matters. Just as love is love, life too, is life.

As Martin Luther King Jr. so famously said:[287]

"Injustice anywhere is a threat to justice everywhere."

Reflection:

What does justice mean to you?

Example from Marilyn: Justice means every life form is treated fairly and can live without abuse, in peace, with freedom, and with the potential to be the best they can be.

The Elephant-Sized Outrage in the Room

We learned a lot from Cecil the lion. We learned that outrage is not just a social media phenomenon; it can have consequences. Walter J. Palmer, Cecil's hunter, went into hiding, after all.[288] We learned that major U.S. airlines were carrying animal trophies across the world. They have since announced the termination of this activity.[289] Consumer attention can be effective when we unite around a cause.

We learned that the animal rights community was disgusted with the outrage over Cecil, because billions of factory-farmed animals are slaughtered in the U.S. annually. "Where's the outrage for what's on our plates, in our refrigerators, and in our bellies?" they asked.

Sorely missing from the discourse on animal abuse is the consideration that chickens, cows, turkeys, sheep, and pigs are intelligent beings who live in complex societies.[290] Their murder is not justifiable. And, if we are upset about Cecil, we need to be upset about these billions of "Cecils" too.

In the previous essay, I wrote about making the connections between Cecil and other justice movements (including animal rights, black rights, women's rights, and LGBTQ rights), because a united movement would be more effective. If we want to save animals yet take a stance against the veganism movement, we are shooting ourselves in the foot. If we say that black lives matter too but not other human or animal lives,[291] we aren't supporting our brothers and sisters in different aspects of the same fight; instead we are choosing fear of losing attention and priority given to black lives. Yes, black lives matter, too. Period. Yes, animal lives matter too, period. And so on.

A divisive justice movement devolves into a game of my outrage is better than yours,[292] and fights for what is perceived to be a limited pie of justice. The truth is that justice is connected to all of life, and every effort toward justice supports every other effort in the same direction.

It's important to be consistent as activists, heck, as human beings! It's time to acknowledge and act to remove the elephant-sized outrage in the room with all social justice activists: the genocide of factory farmed animals. And, in the spirit of unity with all of our fellow citizens, it's time to start practicing the same concern and compassion for all humans and other animals in our quest for justice.

Reflection:

How do you think justice-related social movements can unite and work more closely together?

Example from Marilyn: I think they need an umbrella organization or platform that gives equal weight to all the priorities, gathers funding and other resources for each group, and explicitly connects the dots between each priority.

Chapter 5 Notes

---

[250] Read more about Kantian ethics here:
http://www.csus.edu/indiv/g/gaskilld/ethics/kantian%20ethics.htm

[251] Learn more about the movie: http://www.thegoodliemovie.com/

[252] Join Climate Healers via their Facebook group: https://www.facebook.com/climatehealers

[253] Read the full paper here:
http://www.worldwatch.org/files/pdf/Livestock%20and%20Climate%20Change.pdf

[254] Learn more about mycelial networks here: http://podcasts.personallifemedia.com/podcasts/224-living-green/episodes/2914-paul-stamets-fungal-intelligence-21st

[255] Watch a PBS video here about how plants communicate: http://www.pbs.org/video/2338524490/

[256] Learn more about the documentary Cowspiracy here: http://www.cowspiracy.com/ and you can also watch it on Netflix.

[257] The book, Empathic Civilization, really opened my eyes about how much we are wired for empathy:
https://www.goodreads.com/book/show/7502392-the-empathic-civilization

[258] Learn more about this conference: http://www.wipce2017.com/

[259] Read this article about how long racism has been around: https://newrepublic.com/article/122073/our-racial-history-isnt-back-haunt-us-it-never-left-us?utm_source=Sailthru&utm_medium=email&utm_term=TNR+Daily+Newsletter&utm_campaign=Daily+Newsletter+-+6%2F19%2F15

[260] See the previous essay in this chapter.

[261] See what the Department of Justice has compiled in terms of racism examples in Ferguson:
https://thinkprogress.org/9-egregious-examples-of-racism-in-ferguson-uncovered-by-the-department-of-justice-69625ee29b9f

[262] Learn more about what happened in Baltimore:
https://www.nytimes.com/2015/05/10/opinion/sunday/how-racism-doomed-baltimore.html

[263] Read New York Times' coverage of Dylann Roof's murders:
https://www.nytimes.com/2015/06/20/us/charleston-shooting-dylann-storm-roof.html

[264] Read more about historic white supremacy:
https://www.theatlantic.com/politics/archive/2015/06/take-down-the-confederate-flag-now/396290/

[265] Learn more about Rachel Dolezal: https://www.nytimes.com/2015/06/17/us/rachel-dolezal-nbc-today-show.html

[266] Read the Forbes review of the movie "Black or White:" https://www.forbes.com/sites/rebeccatheodore/2015/01/30/black-or-white-movie-review-race/#66c3ca89512f

[267] Read more about how all men are genetically related: https://www.newscientist.com/article/dn23240-the-father-of-all-men-is-340000-years-old#.VYmas_lVhBc

[268] Read more about human lineages: http://www.nature.com/ejhg/journal/v22/n9/abs/ejhg2013303a.html

[269] Learn more about Spencer Wells: http://www.nationalgeographic.com/explorers/bios/spencer-wells/

[270] Read the Guardian's review of The Journey of Man: https://www.theguardian.com/books/2002/nov/23/highereducation.scienceandnature

[271] Learn more about the Genographic Project: https://genographic.nationalgeographic.com/

[272] See Chapter 11 for more on these science-based tools, and about leading with integrity.

[273] See the "Seeing Beyond Orange…" essay series also in this chapter.

[274] Learn more about the San Francisco County Jail: http://www.sfsheriff.com/jail_info.html

[275] Learn more about this type of jail structure: http://www.ls2.soziologie.uni-muenchen.de/downloads/wise0708/space/panopticon.pdf

[276] AVP's website: http://avpcalifornia.org/

[277] Ibid.

[278] See the essay entitled, "(No) Scientific Basis For Racism" also in this chapter.

[279] Read more on the New York Times' take on Internet Vigilantism: https://www.nytimes.com/2015/07/30/us/cecil-the-lion-walter-palmer.html

[280] Read the blog here: https://thinkprogress.org/the-science-of-why-you-are-so-upset-about-cecil-the-lion-2639a26c7467

[281] Read more about this perspective here: https://blavity.com/maybe-you-should-care-about-black-people-as-much-as-you-care-about-cecil-the-lion/

[282] Read other quotes from Paul Farmer: https://www.goodreads.com/author/quotes/6684.Paul_Farmer

[283] Read more about how to connect the dots from one of my heroes, Dr. Will Tuttle: http://thethinkingvegan.com/interviews/will-tuttle-circles-compassion-connecting-issues-justice/

[284] Learn more about this incredibly important book: http://veganpublishers.com/multimedia-archive/injustice-anywhere-essays-connecting-human-animal-and-environmental-well-being/

[285] See Chapter 9 for more about motivation, and Chapter 11 for how to integrate science-based tools.

[286] Learn about how the LGBTQ movement succeeded: http://newpol.org/content/us-gay-rights-movement-mobilizes-wins-victory-against-discrimination

[287] Read the full quote here: http://www.goodreads.com/quotes/631479-injustice-anywhere-is-a-threat-to-justice-everywhere-we-are

[288] Read more about Walter J. Palmer's hiding: http://www.cnn.com/2015/07/30/us/walter-palmer-whereabouts/

[289] Learn more about the airlines that were transporting animal trophies: http://www.reuters.com/article/us-zimbabwe-wildlife-airlines-idUSKCN0Q90KT20150804

[290] Learn more about these intelligent animals: http://www.upc-online.org/thinking/chicken_human_relationships.pdf

[291] Read more about why black lives matter: http://fusion.kinja.com/the-next-time-someone-says-all-lives-matter-show-them-1793849332

[292] Read more about how the tendency humans have to compete over what's more of an outrage: https://www.theatlantic.com/entertainment/archive/2015/07/outrage-rip-cecil-lion/400037/

Chapter 6: Meditation - Your Super Power

Meditation Goes Mainstream

At Alchemus Prime, we are often asked why meditation is part of our model. Our answer is typically something like this: leading thinkers across generations, including Steve Jobs[293] and Einstein,[294] attribute their creative leaps to their meditation practice. In our own experience, meditation is too important not to bring to our clients. Meditation has powerful benefits.

These days, meditation is being applied in various settings. We apply it in our workshops and retreats. In one of our retreats, we took our client through a daily meditation for all 4 days. This leader began to see the correlation between meditating and a decrease in stress levels; research[295] confirms that experience. Our client also reduced the tendency to overthink and be anxious; science also supports this finding.[296]

Recently mindfulness meditation has become popular in schools, with promising results. The Mindful Life Project teaches mindfulness meditation[297] at the West Contra Costa Unified School District on a weekly basis to third graders. There appears to be a correlation between meditation and a drop in behavioral issues and suspensions in the school.

Visitacion Valley middle school in San Francisco, a very troubled school, is also benefiting from a meditation program called Quite Time,[298] which uses Transcendental Meditation (TM), a technique I swore by and practiced twice daily until I learned Vipassana and switched camps.[299] Here, the results are stunning, according to this article from The Guardian:[300]

"In the first year of Quiet Time suspensions at Visitacion Valley – which has 500 students aged 11-13 – were reduced by 45% (pdf).[301] By 2009-10, attendance rates were over 98% (some of the highest in the city), and today 20% of graduates are admitted to the highly academic Lowell High School[302] – before it was rare for even one student to be accepted. Perhaps even more remarkable, last year's California Healthy Kids Survey[303] from the state's education department found that students at Visitacion Valley middle school were the happiest in the whole of San Francisco."[304]

It's time to pay attention to this powerful method for addressing challenges at the root level. Meditation allows us to connect to our inner selves, and to acknowledge and release pain and trauma. This makes room for the true self to emerge, and to thrive. Leading from the true self, we become aligned with our purpose, and eventually, unstoppable.[305] It's heartening to see meditation becoming popular in the workplace and in schools.

Reflection:

Describe your meditation practice, if you have one. If you do not, what kind of meditation practice would you like to try out? Why?

Example from Marilyn: I meditate using the Vipassana technique, which entails sitting quietly with eyes closed, and scanning the body in a uniform way to observe sensations. Each sensation has to do with a positive or negative experience, and by noticing them, I can detach from them.

Workplace Benefits of Mindfulness

Ever feel like your mind is wandering? Well, you're not alone. I learned mindfulness techniques a few years ago and apply them in my work at Alchemus Prime, applying our science-based model.[306]

A recent blog on mindfulness featured by The London School of Economics' Business Review[307] describes much of the research on the benefits of mindfulness practices in professional settings. I'll summarize it using excerpts. Why? Because the research is so delicious to read and I can't say it better. Mindfulness means being in the present moment with our full, non-judgmental attention. This is a non-trivial endeavor, but becomes easier with practice over time.

Meditation can help us be more productive by reducing the wandering of the mind:[308]

…research suggests your mind wanders about 50% of the time[309]…Meditation lowers activity[310] in the brain region responsible for attention lapses, cutting these by up to 50%.[311]

Mindfulness can help us make better decisions:[312]

Mindfulness guards against the unconscious reactions that produce irrational decisions. Indeed, meditators make more rational decisions and avoid common decision errors, such as continuing to spend money on losing projects[313] or reacting emotionally[314] to unfair situations.

Amazingly, mindfulness can lengthen and enhance our lives and careers:[315]

Internationally-recognized research teams have shown mindfulness slowing biochemical aging, including reducing inflammation[316] and preserving DNA health.[317] Other studies show mindfulness may strengthen disease resistance[318] and immune system functioning.[319] Remarkably, mindfulness may even slow the decay of brain tissue[320] connections from aging. Like a road with fewer potholes, meditators have younger-looking brains[321] with fewer gaps between neurons that may interrupt effective focus and thought.

I grew up in Fiji where the roads have many potholes, so I'll take a road with fewer of them, and a brain with fewer focus gaps, any day! In a nutshell, you and your team cannot afford not to meditate. Start a mindfulness practice today and reap the rewards for the rest of your career and life.

Reflection:

How might you promote mindfulness in your workplace?

Example from Marilyn: I lead guided mindfulness sessions for colleagues and clients. For example, they listen to my guidance as they focus on breath, sound, or body sensations.

Mindfulness Improves Focus

If you're eating, texting, reading, or browsing Facebook or Twitter while listening to music, or any combination of these, you suffer from the effects of multitasking: distraction, racing thoughts, inability to concentrate, and lack of productivity. What's worse, these effects can last even after we stop multitasking.

Recent research[322] suggests that mindful breathing improves focus. Heavy multitaskers improved their focus more than light multitaskers after they performed a mindful breathing exercise prior to taking attention-measuring tests. While the study[323] does not measure the long-term effects of mindful breathing on multitaskers, it offers a good start in that direction with an initial short-term investigation.

As always, it's important to remember what meditation is, and to avoid framing your practice as a failure when your mind wanders. As Prof. Green, one of the study authors describes, mindfulness is a practice:[324]

"When you notice your attention slipping away, you bring it back over and over. You're practicing that skill, refocusing your attention."

Of course, the most important ingredient in our mindfulness practice is non-judgment. At Alchemus Prime, we incorporate mindfulness into our retreats and workshops, offering guided opportunities to start or build the habit of being present in every moment, which offers many benefits, including increasing our grey matter and reducing irrationality and poor decision-making.[325]

Ultimately, the benefits of mindfulness are so valuable that it makes no sense not to practice mindfulness meditation; mindful breathing is just one method. The proof is in the practice.

Reflection:

Reflect on how you felt the last time you were multitasking, and compare it with the last time you did one thing at a time.

Example from Marilyn: I was listening to music while eating breakfast and drawing my Zentangle. I felt scattered. When I finished eating and drawing, I decided to turn the music off and focus on work. Mentally, I felt much clearer.

Solitude, Rest, Meditation, and Wellness

If you're an introvert like me, you'll understand the importance of being alone to recharge your batteries. I love being social, including networking, facilitating workshops and retreats, and dancing, but when it's all over, I need time alone to rest and rejuvenate.

Even if you're not an introvert, research[326] suggests that being alone is conducive to resting: Hubbub surveyed 18,000 people in 134 countries, and found that being alone was the number 3 most restful activity. The number 1 activity was reading, which is also often carried out alone. Number 2 was being outside in nature, which is proven to be good for you.[327]

Back to being alone – it turns out that women and millennials chose this option most frequently. The Hubbub study[328] also showed that young people and men preferred listening to music, which came in at number 4, while number 5 was doing nothing in particular, which led to guilt in the 31-45 age bracket – any workaholics out there?

If you have a meditation habit like me, you might find that you feel rested and energized after your meditations. I do Vipassana meditation for one hour twice per day, and often emerge from those sessions feeling like I've just had a deep, restful nap. Interestingly, meditation appeared at number 10 on the list, perhaps because many of us don't do it.[329]

How much rest is enough? The Hubbub study showed a peak of 5-6 hours,[330] which is proving true for me these days. The bottom line in all this, of course, is to achieve an optimal level of wellness and productivity.[331] Wellness[332] forms the foundation for all we do, whether it's work or play. When we're healthy, we perform more productively at work, have better relationships, and have a better outlook on life. In a climate-disrupted world where our every life-affirming action counts, wellness is the key to driving sustained behavior change.

Reflection:

How do you best recharge when you're tired? Is it alone or with people?

Example from Margaret: I like to be alone. I typically meditate, listen to music, or relax with a book or sewing project.

The Biomimicry of Vipassana

Recently I participated in a ten-day Vipassana Meditation Training in the California desert.[333] It had been on my bucket list for over a decade, so I figured it was time. Having practiced Zen meditation, guided visualization, Reiki, Transcendental Meditation (TM), and Mindfulness-Based Stress Reduction (MBSR),[334] I still felt something was missing from my meditation jigsaw puzzle.

Well, I was right.

Vipassana has made my practice whole. Paraphrasing S. N. Goenka, whose recorded video and audio instructions we followed in the course: Vipassana is about connecting to one's own direct experience of one's own reality from moment to moment. This is different from relying on a deity, a mantra, a visualization or other vehicle to reach deeper consciousness. The result of being in the present moment observing one's own reality via sensations on the body is that deep mental aversions and cravings have the opportunity to surface and evaporate.

Amazing, right? What was missing for me was the complete, pristine technique, as taught by Siddhartha Gautama, the Buddha, for accessing and cleaning the subconscious mind. Just as we feel satisfied when we brush our teeth or clean our space or dishes, cleaning the mind is very beneficial and leaves us open to creativity and productivity.

There's more. I was fascinated by Goenka Ji's[335] discourses each evening, especially because he gave examples from Nature. As a student of biomimicry, my ears immediately perked up. A few themes stood out that I'll summarize:

1. **Life affirms life.** Biomimicry is about emulating nature's principles and processes. One overarching principle of nature is that it creates conditions conducive to life.[336] In Vipassana, the gem of Sila (Morality), contains the principle of cause and effect, which stipulates that if you do something negative (steal, lie, etc.) it will have a negative, and immediate effect. Therefore the focus is on honesty, compassion, love, peace, and happiness in all actions to ensure good karma.

2. **Direct observation.** The Scoping phase of biomimicry[337] involves deep observation of examples in nature that can inform design principles for creating solutions for myriad social, engineering, and

other solutions. In Vipassana, we deeply observe our own bodies for sensations, which signal what present and past miseries we are carrying around with us.

3. **The four elements.** Earth, air, water, and fire make up our natural world and models from nature show how these elements can be used to advance life, such as using wind, fire, or water for seed dispersal. During the Vipassana course, I learned that the sensations we feel on the body correlate to the four elements: fire shows up as heat or cold and connotes anger; air manifests in the body as movement, such as trembling or butterflies in the stomach; earth is either a heavy feeling or a feeling of lightness, such as after a large or small meal respectively; and water has to do with cohesion or feeling like things are coming together in the body, whether positive or negative.

Goenka Ji also used a few metaphors from nature to help explain how the human mind functions. I found these interesting:

• **Multiplication:** in nature, seeds follow a principle of multiplication to spread their DNA. Every seed sprouts into a full plant or tree with a lot of fruit, which in turn all have many seeds. This is how aversions and cravings multiply in the mind – one seed or thought is planted, for example: "This is going to be a really tough week," and then multiple thoughts are generated on top of this one, leading to overthinking, anxiety, worry, fear, and stress.

• **Reserves:** A body stores fat and other nutrients, and when one starves or fasts, these reserves are brought into use to help keep the body alive. If no food is taken, eventually the person dies. Similarly, when aversions and cravings are simply observed with equanimity, they evaporate and old, deep traumas surface. If those are also only observed, not added to, eventually they are released too, and this is how one reaches a state of liberation.

I found these commonalities between biomimicry and Vipassana very intriguing, especially for my quest to bridge personal and planetary wellness[338]…what we do to ourselves. we are doing to the world, and vice versa. The sooner we understand this, the better we can create harmony between humans and the rest of nature.

Reflection:

Consider your own breath: it connects your inner world to the outer world. For a few minutes, simply notice your breath, and then write down what you observe.

Example from Marilyn: I notice my breathing is shallow when I'm at rest, and there's some warmth on my nostril edges when I breathe out.

Beyond Meditation: Cultivating Mindfulness

As I reflect on my 12-year journey of learning different meditation techniques, the latest being Vipassana,[339] I am reminded that the benefits are tremendous, both based on research[340] and my practice. An article from Fulfillment Daily[341] seems to be describing my life as it lists thirteen ways to cultivate mindfulness, backed by research: check it out. I'll summarize those thirteen practices here with my own examples:

1. **Walking:** I walk for an hour or two daily, often on the beach. The sense of peace and inspiration is rejuvenating.

2. **Daily tasks:** I find tasks like cleaning very satisfying when I can be in the moment, with full awareness of what I'm doing.

3. **Creativity:** In my Vipassana retreat, so many ideas came into my mind; I am currently working on four books and developing a new wellness framework.

4. **Breathing:** Being aware of my breath is a powerful way to focus on the present moment whenever I find myself stuck in the past or worrying about the future.

5. **Unitasking:** I am much more productive when I do one thing at a time. These days I set aside time for one task or a piece of one task, with no distractions. I tend to accomplish more than I expect to, each day.

6. **De-phoning:** I don't check my phone as often as I used to. After ten days of not having my phone with me during Vipassana training, I was reminded that I've often overestimated its importance.

7. **Novelty:** I seek out new experiences through food, movies, books, walks, and art.

8. **Nature:** I spend more and more time outdoors communing with nature. I notice the trees, flowers, birds, waves, and everything else in my path.

9. **Emotions:** I notice what I'm feeling and just pause to feel it, accepting what is instead of suppressing it, which always means I will have to deal with it later. Of course, this is a lifetime of practice, and I'm slowly improving, especially after Vipassana.

10. **Meditation:** I now meditate using the Vipassana technique for two hours per day. My previous regimen was transcendental meditation for 40 minutes per day, which was energizing, but did not include focusing on my own reality.

11. **Inputs:** I am careful to put healthy, organic, plant-based, and gluten-free food into my body, and focus my entertainment choices on topics like comedy, animal protection, nature, foreign and

independent drama, and other non-violent topics. The exceptions of course are Star Wars and Transformers movies and cartoons, which I don't know how to give up!

12. **Laughter:** Watching life's ups and downs, remembering not to take everything so seriously, and using humor are essential. I'm slowly getting better at this.

13. **Imagination:** Allowing the mind to roam and wander is a crucial way to access new ideas. My daydreams always bring me new ideas, stories, and solutions.

This is a lot to digest. If I had to pick one or two of these to emphasize, it would be #8: get outside. Research suggests that spending time outdoors can relieve stress,[342] and improve energy levels,[343] memory and attention.[344]

And, #10: meditate! Mindfulness meditation can change our gene expression[345] to lower our inflammatory response, making us more resilient. The strength of the mind-body connection becomes more apparent as we pause and connect with our own unique realities in the present moment.

Reflection:

What do you think is the most compelling reason for you to meditate? Why?

Example from Marilyn: For me it's the genetic connection – meditation can change our wellbeing at a genetic level – that is so powerful!

Chapter 6 Notes

---

[293] Read more about Steve Jobs and his meditation practice: http://www.businessinsider.com/steve-jobs-zen-meditation-buddhism-2015-1

[294] Read about Einstein and how his views stem from some of the consciousness states achieved through meditation: http://www.tm.org/blog/enlightenment/albert-einstein/

[295] Meditation reduces stress, according to this research paper in the journal *Social Cognitive and Affective Neuroscience*: https://www.ncbi.nlm.nih.gov/pmc/articles/PMC2840837/#B42

[296] Research shows meditation improves brain health, reducing anxiety and overthinking: https://www.davidwolfe.com/meditation-rebuilds-brains-gray-matter-in-8-weeks/

[297] Read more about this school program: http://www.eastbaytimes.com/2015/11/01/richmond-group-brings-mindfulness-to-classrooms/

[298] Read more about how Visitacion Valley Middle School uses Quiet Time for meditation: https://visitacion-sfusd-ca.schoolloop.com/quiet_time

[299] I think you should try different meditation techniques and decide for yourself. I like both Transcendental Meditation and Vipassana. TM helped me a lot with becoming more creative and productive, but the power of Vipassana in cleaning the mind of negative energy has proven too valuable to not undertake twice daily. I can feel the change in terms of how much happier I am, how little external events bother me, and how much more productive and creative I have become.

[300] Read the full article here: http://www.theguardian.com/teacher-network/2015/nov/24/san-franciscos-toughest-schools-transformed-meditation

[301] Read the article here: http://cwae.org/media/SF_Chron_1-13-14-4.pdf

[302] See the high school's website for more about how they operate: https://lhs-sfusd-ca.schoolloop.com/cms/page_view?d=x&piid=&vpid=1239686263777

[303] Read more about the features of the survey here: http://chks.wested.org/

[304] Read more about these "happiest" children in San Francisco: https://www.mnn.com/health/fitness-well-being/blogs/meditation-works-formerly-low-performing-high-school-now-has

[305] See Chapter 11 for more about what we mean by the true self.

[306] See Chapter 11 for more about the Alchemus Prime and our science-based Diamond Model.

[307] Read the full blog post here: http://blogs.lse.ac.uk/businessreview/2016/03/02/mindfulness-has-big-impacts-for-performance-decision-making-and-career-longevity/

[308] Ibid.

[309] Read the full paper here:
http://www.danielgilbert.com/KILLINGSWORTH%20&%20GILBERT%20(2010).pdf

[310] Read the full paper here: http://www.pnas.org/content/108/50/20254.long

[311] Read the paper summarizing the studies here:
https://labs.psych.ucsb.edu/schooler/jonathan/sites/labs.psych.ucsb.edu.schooler.jonathan/files/pubs/mindulness_and_mind-wandering.pdf

[312] See Note #242.

[313] Meditation can de-bias the mind, read more about the mechanisms here:
http://www.truedevelopment.se/Debiasing%20the%20mind%20through%20meditation%20-%20%20mindfulness%20and%20the%20sunk-cost%20effect.pdf

[314] Read the paper on how meditation reduces emotional reactions and improves rational decision-making:
http://journal.frontiersin.org/article/10.3389/fnins.2011.00049/full

[315] See Note #242.

[316] Read more about how meditation lowers inflammation:
https://www.ncbi.nlm.nih.gov/pmc/articles/PMC4039194/

[317] Mindfulness improves longevity at a DNA level: http://www.psyneuen-journal.com/article/S0306-4530(13)00453-8/abstract

[318] Meditation can increase our resistance to diseases:
https://www.ncbi.nlm.nih.gov/pmc/articles/PMC2725018/

[319] Meditation can strengthen the functioning of our immune systems:
http://centerhealthyminds.org/assets/files-publications/DavidsonAlterationsPsychosomaticMedicine.pdf

[320] Meditation can slow the rate of brain tissue decay:
http://journal.frontiersin.org/article/10.3389/fpsyg.2014.01551/full

[321] Meditators have younger brains with more fluid intelligence:
https://www.ncbi.nlm.nih.gov/pmc/articles/PMC4001007/

[322] Read more here on mindfulness and improved focus:
https://www.sciencedaily.com/releases/2016/04/160419081758.htm

[323] Access the scientific study here: http://www.nature.com/articles/srep24542

[324] See Note #257.

[325] See the previous essay for the workplace benefits of mindfulness.

[326] See the full results of the Hubbub Study: http://hubbubresearch.org/rest-test-results/

[327] See my blog about how children become smarter and healthier when they spend time in nature: http://www.integritusprime.com/nature-makes-children-smarter-healthier/

[328] See Note #261.

[329] See the essay entitled, "Meditation Goes Mainstream" also in this chapter. It's time more of us joined the meditation bandwagon, simply because the benefits are phenomenal.

[330] See the BBC article that reports on the Hubbub study: http://www.bbc.com/news/magazine-37444982

[331] Some habits that boost productivity: http://www.integritusprime.com/10-habits-that-boost-productivity/

[332] Chapter 2 covers wellness and diet in detail.

[333] Learn more about Vipassana: https://www.dhamma.org/en-US/index

[334] See Note #234.

[335] "Ji" is a term of respect, like "sir" in Hindi, and often used for teachers and gurus.

[336] Creating conditions in which life thrives is a basic principle of Biomimicry. Read more about it: https://biomimicry.org/learning-nature-designing-nature-regenerative-cultures-create-conditions-conducive-life/

[337] For more on Scoping in biomimicry (which means to emulate nature), see the essay entitled, "Pioneering a Harmonious Future: Biomimicry for Social Innovation" in Chapter 7.

[338] See Chapter 1.

[339] See Note #268.

[340] Meditation has many benefits, see the rest of the essays in this chapter, and this article: http://www.huffingtonpost.com/2013/04/08/mindfulness-meditation-benefits-health_n_3016045.html

[341] Read the full article here: http://www.fulfillmentdaily.com/why-mindfulness-is-more-than-just-meditation/?utm_content=buffer7e416&utm_medium=social&utm_source=facebook.com&utm_campaign=buffer

[342] Being outdoors can alleviate stress; see this article: https://phys.org/news/2012-02-green-spaces-stress-jobless.html

[343] Being outside improves our energy levels: http://www.sciencedirect.com/science/article/pii/S0272494409000838

[344] Being in nature can improve our memory and attention: http://ns.umich.edu/new/releases/6892

[345] Meditation can lower our levels of inflammation, intervening at a genetic level: http://www.huffingtonpost.com/2013/12/09/mindfulness-meditation-gene-expression_n_4391871.html

Chapter 7: How Nature Innovates

Sacredness as Biomimicry

Biomimicry[346] is defined as the intentional emulation of nature's wisdom. As humans transition into a future that promises chaos, biomimicry is an important tool that can bring harmony.

The three essential elements of biomimicry[347] are an ethos of respect and humility, emulating nature's design, and (re)connecting with nature by being outside and observing its patterns for inspiration.

If we extend biomimicry to include emulating not only other species, but also humans who live or have lived in sync with nature, we can look to indigenous people's practices for wisdom and guidance. A social aspect of biomimicry, if you will.

Expanding the ethos element of biomimicry, we might add what we can observe in the practices of native peoples, and what I observed while working in Navajo Nation:

- See yourself as nature, and nature as yourself in your worldview
- Take only what you need
- Use everything you take
- Revere nature, the source of your sustenance
- Perform sacred rituals to practice that reverence

Now, there are many ways to define leadership.[348] One businessman defines leadership behaviorally:

"Leadership is the behavior that brings the future to the present, by envisioning the possible and persuading others to help you make it a reality." -- Matt Barney, founder and CEO, LeaderAmp

This resonates with the vision of my company, Alchemus Prime. Our goal is to empower leaders to safeguard the future....now.

In order to achieve this goal, we incorporate our extended view of biomimicry and its ethos, offering mediation and biomimicry exercises that encourage all three elements of biomimicry: ethos, emulate, and (re)connect. We hope to nurture the realization and practices of *sacredness*. This is not necessarily a religious feeling, but a profound connection and reverence for nature.

Nature is a blessing to us; how might humans be a blessing to nature?

Feeling this sacredness, how might we act? Would we destroy? Or would we protect? Do we exploit for short-term misguided gain, or do we nurture for all future generations to enjoy? This decision marks the course of true leadership, and defines the success of humans. For we cannot destroy nature, the life support system of human lives, and indeed of all life, and hope to persist. We must rediscover the joy of sustaining all life as the core business of business, and all human action.

Reflection:

Think about a time you were outside and you felt a oneness with nature. Write a few words to describe how you felt.

Example from Marilyn: I was at the beach and I saw two dolphins, jumping in unison. I felt a sense of euphoria. I felt connected to their harmony of movement. It was a very joyful moment.

Pioneering a Harmonious Future: Biomimicry for Social Innovation

My last thoughts on biomimicry talked about the importance of sacredness as a biomimetic practice.[349] This essay will focus on the power of applying biomimicry to social innovation, or the work we do to leverage relationships, communication, and behavior change in diverse professional settings.

For six days in the spring of 2015, I participated in a Biomimicry Thinking for Social Innovation Immersion Workshop[350] at the Occidental Arts and Ecology Center,[351] led by intrepid instructors Toby Herzlich and Dayna Baumeister. This first-of-its-kind workshop is part of a movement called Biomimicry for Social Innovation;[352] the goal is to shape human communication, cooperation, and action in ways that mimic nature to create conditions conducive to life.

I showed up with my fellow participants and dove into what would become a life-changing, soul-affirming, intense, and beautiful week of learning from nature through deep biological inquiry, fun and challenging teamwork, and many games and exercises that yielded a-ha moments for all twenty-six of us. We observed, designed, learned, shared, presented, performed hilarious skits, danced, laughed, cried, and became humble at the realization of how much more there is to learn and do. We became colleagues, friends, and a community…we found our tribe. Some of the highlights for me are:

Mutualisms

A mutualism is a win-win relationship between two organisms that exchange something different of value with one another so that both can live well. Reinforcing feedback loops keep the relationship positive over time, and this type of relationship is optimal because it takes less energy to participate in, and provides ongoing benefits to both parties. Sounds pretty wise, right? As we cultivate relationships in the workplace and elsewhere, it will behoove us to focus on cooperation, mutually beneficial exchange, and tight feedback loops (i.e. instant and effective communication) so we adjust to changes in our context. And, we can practice applying the really golden rule in nature: *when disturbances and stressors increase, mutualisms increase.* How might we cooperate and help each other more, when times get more difficult?

Scoping (Function and Context)

Scoping is the first phase of a biomimicry inquiry. Function is what we want our design (or solution) to *do*; context is a description of the situation we are operating in. For example, we might want our design to create illumination, and our context might be in poor villages in the developing world. In biomimicry, we would ask, "How does nature illuminate?" Then we would look at champion examples and deep patterns in nature and derive design principles that would guide our solution building.

Adapting to Change

One of the Life's Principles in Biomimicry is called "Adapt to Changing Conditions."[353] This includes incorporating diversity, maintaining integrity, and building resilience to preserve the function of our design or solution, as context changes. An example of resilience I experienced was pretending to be a bird, and moving with my team of about thirteen participants in a flock-like manner on the beach as part of a biomimicry exercise. Our simple rules were: stay equidistant from one another, fly into the wind, disperse as needed then return to formation, and the last rule: follow these rules. We learned that following one of us was flawed, as we would make the same mistakes as the person we were following. Changing the rules, we discovered, could change our configuration sometimes dramatically.

In this exercise, the presence of a predator (picture Toby screaming and charging toward us in a random pattern without warning) would cause us to disperse as we need to (decentralized authority); screech or flap our wings frantically (diversity of communication modes); and then return to our formation when it was safe. If the wind blew in unpredictable ways (indicated by Toby), we would fly into the wind, according to one of our rules. If the predator ate one of us (thankfully this didn't happen, as Toby is nowhere near cannibalistic), the rest of us could still survive and keep moving forward because we all have the same survival skills (redundancy). Clearly, this was a hilarious and simultaneously essential exercise to sensitize us to how nature builds resilience through decentralization, diversity, redundancy, simple rules, and the overarching mantra: cooperate, cooperate, cooperate!

As I reflected on this exercise, I realized that innovative organizations are embracing these principles of resilience. For instance, Google recently launched a supplier diversity program,[354] and they have decentralized purchasing, which allows every employee to make their own purchasing decisions; this has fewer and smaller repercussions if problems arise with any single purchase or supplier, compared to a centralized model.

What these gems and all my learning in this workshop means for me, and for the Alchemus Prime Diamond Model,[355] is that the more biomimicry I infuse into the model, the more integrity the model contains, because the central goal of the model is to actualize win-win solutions for all life. Just as importantly, I realized during this workshop that each component of the model is a mutually-reinforcing partner to the others– for example, biomimicry and design thinking share convergent and divergent processes, and all components speak to the importance of changing behavior to reach our goals. Biomimicry and meditation relate to planetary and personal integrity, and my goal is to merge those for a life-affirming outcome every time the model is applied.

What more affirmation could I ask for? I am on my way, feeling more hopeful and inspired than ever, and grateful for the stellar community of diverse, decentralized, and focused leaders with whom I have the privilege of co-creating a thriving and harmonious future. Let's do this!

Reflection:

Describe a mutualistic or cooperative relationship in your life in which you both win.

Example from Marilyn: Mom and I are very cooperative. She preps ingredients and washes dishes; I cook and bake. Sometimes we switch roles. If there's ever any disharmony between us, we communicate immediately and more than usual until it's resolved.

Biologically, We are Nature

In a previous essay, I wrote about mutualisms[356] – cooperative relationships – that govern the way organisms in nature behave in response to change. When disturbances happen in nature, such as earthquakes, floods, or fires, mutualisms increase. Essentially, when nature experiences stress, she focuses more on relationships, cooperation, and communication.

Well, guess what? So do we.

Research[357] indicates that we humans respond to stress with a "tend-and-befriend" tactic,[358] which reduces fear and increases optimism through social caregiving (via the biological hormone oxytocin), a reward system (via dopamine), and attunement or heightened intuition and self-control (via serotonin). Perhaps even more importantly, whenever we choose to help others, we activate this state, which Kelly McGonigal calls the "biology of courage."[359]

Now, if nature and humans are wired to behave in exactly the same way when responding to stress, doesn't that make humans and nature…the same? Similar, at least.

One of my big questions right now is how we might expand biomimicry to include emulating human practices that are in tune with nature, such as the practices of indigenous peoples living in subsistence. I call this "social biomimicry." Another potential arena of interest is "human biomimicry:" how human biology is wired to behave in the same ways that the rest of nature behaves in response to change and stress, and how we might learn from these similarities.

As one of the Life's Principles states, we must "Evolve to Survive."[360] This is especially true in this time of profound change and uncertainty. How might we look more deeply at nature by going outside, and by going inside through biological and meditative inquiries to understand and adapt our practices to be in harmony with the Earth? That is, to me, one of the most important questions of our time.

Are you stressed? Try reaching out to others. Connect. Communicate. Our strength grows through our relationships.

Reflection:

How do you typically respond to stress? How might you change that response to do more of what nature does – cooperate and communicate?

Example from Marilyn: Typically, when stressed, I take alone time but also need to resolve the issue in a timely way. I already communicate a lot but become frustrated when the other party won't reciprocate. I could benefit from practicing more patience.

Maintaining Integrity Through Change

In this essay I'll discuss two examples from nature that help describe what I do with clients and why it's so rewarding to do this work. In a nutshell, Alchemus Prime helps clients maintain integrity while adapting to profound changes.

In a world that is racing by us with new technologies and mass species extinctions, sometimes it's difficult to get a grip on who we are, why we're here, and what we can best be doing to make a difference. These two examples may be instructive and inspiring:

Metamorphosis: Self-Renewal

When a caterpillar goes into its cocoon, it takes a leap of faith, trusting that things will be all right despite profound change in its form and function. As a leader facing uncertainty and paradigmatic change, you might ask yourself what you need to take with you on this journey. I did ask this very question a few years ago at the National Bioneers Conference[361] in a workshop the brilliant Toby Herzlich[362] was leading. My answer was the "Maintain Integrity through Self-Renewal" card from the Life's Principles Leadership Deck.

My epiphany at that time was that everything that I need to maintain my integrity as a leader and environmentalist is what I will take with me through profound change; everything else I will leave behind. For me, this meant my values, principles, passion, and self-care, were essential components to take with me. What is essential to your function that you want to take with you, no matter how much change life may bring?

Water: Changing States

Recently it was my birthday and I woke up with an intense insight, as if someone had planted a burning seed in my mind. The insight was this: water can adapt to external change (such as heat or cold) by changing its state, and in some cases this change in state can pollute or purify the water, but water never loses its integrity and identity as water. I lay in bed pondering and digesting this for a while, and wrote a poem about it as well:

water

May 4, 2015

i am water

i adapt to change

i evolve to survive

i am living

at the edges

of being alive

embodying life's principles

carrying the blueprint

of life's drive

when i am contaminated

i change my state

i evaporate

into vapor, rising

condensing somewhere else

i return

to my true self

you don't stop

calling me water

when i am vapor

or ice

because even though

i change my state

i know why and when

and for how long

and that must suffice

my integrity

is intact

my identity

unchanged

i change states

to remain

the same

to purify

and return

always

to my pure

form

i am in life

and death

and in that

which is reborn

i am consistent

even as i yield

to the pressures

of an unbalanced world

that is turbulent

and imprecise

in its disturbances

i find creative solutions

in the surprise

i am steady

and supple

in the midst

of uncertainty

and strife

in the darkness

i flow

trusting

in the coming

of the sunrise

i am water

the changing

and integritus*

essence

of life...

Think about polluted water, say, in a river. When conditions heat up, water evaporates, leaving behind particulate impurities, and transforms into vapor that drifts off to another location, eventually to fall as nourishing rain. Water can also freeze and create a layer of ice, upon which snow falls and stays clean and pure. Of course, there are exceptions, such as acid rain,[363] so it's important to know how far you can take this example. In human terms, acid rain is a compromised state, in which we have lost our integrity, and become "polluted."

Water's lesson for me is that I can choose to change in substantial ways to adapt to changing conditions without losing my identity or integrity. In fact, changing myself in deep ways could be advantageous to me, as it can be for water. I can embrace change as an ally, a partner, a friend. How do you see yourself adapting to change: is it making you more yourself?

*Integritus is a word I made up; it means "having integrity." It's part of the name of my blog, Integritus Prime.

Reflection:

Describe a time when your integrity was challenged, and how you responded.

Example from Marilyn: I was in a situation where I was asked to suppress the team's weaknesses, but this felt wrong to me because I saw these weaknesses as opportunities to improve project implementation in the next year, so I prepared a framework document focusing on opportunities and suggestions for improvement, leveraging team members' existing skills and highlighting hiring and training needs.

Bruce Lee's Biomimicry

On the Alchemus Prime website,[364] we have a note about how we serve clients:

"We help our clients become more like water: accepting, adapting to, and flowing with change in ways that maintain and strengthen their integrity while wearing away even the largest barriers."

This note stems from our deep foundations in biomimicry,[365] the art of emulating nature's design and principles. Water is a beautiful example of how nature can teach us to live and function. Here's how I see it: water changes its form without changing its identity: vapor and ice are forms of water and can return to liquid form when circumstances change. Water is soft and flowing, but can wear away the hardest rocks over time. Water surrenders to pressure, gravity, temperature, and other impacts. Its strength lies in accepting change, not fighting against it.

As someone who works to understand and leverage behavior change, this is a truly quintessential example for me to work towards, both personally and professionally.

The Tao philosophy teaches a similar perspective. My favorite translation of *Tao Te Ching* is by Stephen Mitchell; his version of the chapter on water states:[366]

Nothing in the world
is as soft and yielding as water.

Yet for dissolving the hard and inflexible,
nothing can surpass it.

The soft overcomes the hard;
the gentle overcomes the rigid.

Everyone knows this is true,
but few can put it into practice.

Therefore the Master remains
serene in the midst of sorrow,

Evil cannot enter his heart.

Because he has given up helping,
he is people's greatest help.

True words seem paradoxical.

As it turns out, the late martial artist, instructor, actor, director, and legend Bruce Lee also thought deeply about water as a teacher for his life and actions. He embraced the Tao philosophy, and spoke eloquently about his deep epiphanies inspired by water:[367]

"After spending many hours meditating and practicing, I gave up and went sailing alone in a junk. On the sea I thought of all my past training and got mad at myself and punched the water! Right then — at that moment — a thought suddenly struck me; was not this water the very essence of gung fu? Hadn't this water just now illustrated to me the principle of gung fu? I struck it but it did not suffer hurt. Again I struck it with all of my might — yet it was not wounded! I then tried to grasp a handful of it but this proved impossible. This water, the softest substance in the world, which could be contained in the smallest jar, only seemed weak. In reality, it could penetrate the hardest substance in the world. That was it! I wanted to be like the nature of water."

He goes on:[368]

Suddenly a bird flew by and cast its reflection on the water. Right then I was absorbing myself with the lesson of the water, another mystic sense of hidden meaning revealed itself to me; should not the thoughts and emotions I had when in front of an opponent pass like the reflection of the birds flying over the water? This was exactly what Professor Yip meant by being detached — not being without emotion or feeling, but being one in whom feeling was not sticky or blocked. Therefore in order to control myself I must first accept myself by going with and not against my nature.

Part of that last line bears repeating: "…I must first accept myself by going with and not against my nature." I would extend this to say that flowing with our own nature (what we call the true self),[369] is also going with Mother Nature, because we are part of nature and she is part of us.[370] I explore this concept of oneness in Chapter 2 of Earth Champions, my book about true self and leadership.[371]

A business-oriented translation[372] of the same Tao Te Ching chapter on water speaks volumes about how we can hone our behavior:

1. The highest good is like water.

2. Water gives life to the ten thousand things and does not strive.

3. It flows in places men reject and so is like the Tao.

4. In dwelling, be close to the land.

5. In meditation, go deep in the heart.

6. In dealing with others, be gentle and kind.

7. In speech, be true.

8. In ruling, be just.

9. In business, be competent.

10. In action, watch the timing.

11. No fight: no blame.

We aim to craft business initiatives that are in harmony with nature. It's affirming to see the resonance between our approach, Bruce Lee's magic, and the Tao philosophy. It gives me hope that we are converging on what works personally and professionally: embracing and leveraging change to achieve our goals in concert with our own nature and Mother Nature.

The examples of water and Bruce Lee's mastery show us how we can achieve this powerful yin-yang balance between softness and power.[373] As I've said elsewhere, vulnerability is a strength.[374] I'm not saying embracing change is easy, but it's a practice worth adopting, and maintaining with humility and perseverance as we consciously evolve into resilient leaders in a constantly changing world.

Reflection:

Write about one attribute or quality of water that interests you. Why?

Example from Marilyn: Changing of states fascinates me – water, vapor, and ice are so different and yet the identity of water is retained. Is this mastery over change?

Future Cities to be Biomimetic

Cambridge University bioengineer Michelle Oyen[375] is studying how to make future cities biomimetic. Her work[376] is focused on the mechanical properties of biological materials. One of her current projects is looking at how emulating the structure and mechanics of eggshells and bones might yield insights for how to design buildings and entire cities to be more resilient and sustainable.

Oyen's work is aimed at understanding how nature makes building materials at room temperature, without the need for large amounts of energy. Man-made materials are much more energy-intensive; concrete, for instance, accounts for up to 10% of greenhouse gas emissions.[377]

Oyen's approach is part of the budding science of biomimicry, and it can have important implications for the built environment:[378]

"'The natural world and ecological system are maybe the best picture for what a sustainable world looks and performs like,'" says Erin Rovalo, a senior principal of design at the consulting firm Biomimicry 3.8. "'And if our built environment can function like these ecosystems, maybe that's the pinnacle of what sustainable design can be.'"

Another example of how we might design sustainable cities inspired by nature, is with respect to transportation. German neuroscientist Arndt Pechstein explains how proteins carry their cargo within human cells:[379]

"'One motor protein is not only able to carry one specific cargo, so it can be decoupled from that cargo and can take on another," Pechstein says. "You also have different motor proteins that can switch between different infrastructures without any waiting time or delay.'

"And motor proteins communicate with one another, signaling what they're carrying and where they need to drop it off, preventing traffic jams in our cells. That's far more efficient than what happens in our cities, where cars sit idle most of the time and where rush hour can lead to grueling gridlock.

"In studying the pattern that facilitates this sort of efficiency, Pechstein and his team redesigned the car as a concept, and dubbed it Flywheel. In this new, more modular design, the car is round and the passenger cabin seats two people. To seat more people or carry other cargo, it can merge with other vehicles to form a "train." And instead of taking up space in the city, the team proposed an infrastructure design in which the cars—like motor proteins—can switch between being on the road and underground.

"If we can reduce the amount of transit on the surface by guiding it somewhere else, then we would dramatically reduce the noise, pollution, and space required for transport in the city,'" he says.

While this type of inquiry is very exciting and promising, it is costly and will take time; behavior change is also critical, concede Oyen and Pechstein.

Nevertheless, the inspiration is strong in the scientific community to learn from nature and in essence, return to nature in the way we design living systems.

In the words of Pechstein:[380]

"We forget that we are nature, too. And that we are, just as any other species, basically biology, and have something that we can emulate and learn from..."'

Reflection:

What excites you about a possible future where cities mimic nature? Why?

Example from Marilyn: I'm excited about the capacity for net zero waste and energy because then we will truly neutralize our negative impact on our planet. In nature, there is no waste!

Bioneers 2016: Highlights and Questions

Bioneers[381] is, by definition, a recurring utopia. Sometimes it's hard to believe it's a conference!

Everyone is happy: beaming, grateful, present, and kind. For example, a stranger brought hot tea for my friend Katie who was volunteering outside for hours, because she noticed my friend's hands were cold. A co-volunteer, Jake, persistently attended to a disabled man, ensuring that, despite the constraints of the room and AV equipment, that man got to ask his question and receive an answer. This is what the Bioneers community is like.

One of the biggest highlights for me was a mini reunion with my biomimicry tribe: eight out of about twenty students from the inaugural cohort of the Biomimicry for Social Innovation Immersion Course[382] were in attendance. We hugged, lunched, and caught up, basking in the sun and the light of each other's hearts. We supported each other's paths and affirmed our common mission of communication and aligned action to support biomimicry's vital role in addressing environmental challenges.

As I interacted with my cohort members, fellow volunteers and other attendees, I couldn't help but be struck again and again by how much everyone cares. We care about how the person next to us is feeling, about helping someone who can't find a room or building, and about better aligning our work with conditions conducive to life. Nature responds to stress by having each organism extend more help to other organisms. In being kind, attentive, and helpful to each other, we were embodying one of nature's principles. Because, put simply, we are nature.[383] At Bioneers, everyone becomes their true and best self: aligned with life, and focused on transformative solutions for Mother Earth.

Another highlight was meeting new people; amazing leaders drawn to serve the Earth in diverse, creative, and relentlessly just ways. I learned from Kimberle Crenshaw about the disturbing police brutality against black women around the nation,[384] and from Kandi Mossett about the unwarranted way peacefully praying indigenous men, women and children at Standing Rock were treated by police.[385] The audience sat and cried together, feeling the pain of these social injustices in our hearts, and vowing to keep working for a better world.

The inaugural Biomimicry Global Design Challenge Award was presented on the Bioneers stage last year, with Chilean team BioNurse[386] winning the $100K prize. While all the other teams' prototypes were just as stunning in their creativity, BioNurse won due to their strong fidelity to the principles of emulating nature: they modeled their soil health device after the cushion plant,[387] which provides sheltered environments for other species to grow. These young biomimics are a true inspiration…

I also encountered an example of what I call social biomimicry: human practices that are in deep alignment with nature. In Bioneers Cofounder and President, Nina Simon's keynote,[388] I learned about the Okanagan worldview, in which there are four societies: tradition, vision, relationship, and action, which correlate to elders, youth, mothers, and fathers. Tradition refers to indigenous practices, while vision refers to the future. Relationship is the emotional, empathic, and feminine and action is the analytical and masculine. Current societal challenges, according to this worldview, are due to an imbalance caused by too much vision and action, and not enough tradition and relationship. To return to balance, this wise framework asks us to harmonize the masculine and feminine, being and doing, as well as known and new.

Aside from soulful fellowship with my peers; creative, emotional and ethical solidarity; and deep social biomimicry, I also engaged in plenty of scientific and intellectual inquiry, including interesting talks on accelerating solar adoption, and how to sequester carbon in soil. I asked myself some questions that help me think integratively:

- What are the embodied energy implications in the current proliferation of solar photovoltaics?
- What is the role of vegan foods in climate-conscious biomimetic food systems?
- How does nature eat efficiently without polluting?
- How do we heal gender relations and stop violence against people of color, men, women, non-binary and LGBTQ people, children, animals, communities, and entire nations?

Our journey continues: we are all infused with love, hope, and inspiration from this remarkable community, and ready for the next chapter of biomimicry for social innovation: engagement.[389] This was the mantra last year: how to engage people and mobilize them for gender rights, community rights, corporate innovation, creative climate solution building, and more.

Reflection:

Reflect on the Okanagan worldview. How might we bring more empathy, earth-friendly actions, and intuition into leadership?

Example from Marilyn: One very practical way is to improve gender balance in the workforce, especially in leadership positions across the board.

The Biomimetic Life: A Self-Analysis

For a while I've been meaning to analyze my life and work based on Life's Principles[390] from biomimicry, to see how I'm doing in my personal quest to live an ethical life of service to the planet. Some months ago, I attended Barbara Marx Hubbard's Planetary Mission Launch[391] in Berkeley, California, where she spoke eloquently about conscious evolution.[392] She explained that conscious evolution requires three ingredients: consciousness (well, duh), freedom (to co-create), and love (the purpose of evolution). In other words, humans are conscious beings who can impact the course of their future, and the important action to take is to evolve humanity in ways that affirm and love life. Nature is masterful at creating conditions conducive to life. Humans? Not so much.

It's more important now than ever for humanity to align with life. I got started in behavior change through ethics, and have spent sixteen or more years examining my own actions and principles. I know we are all works in progress, which is why it's important to have compassion for ourselves. As I embark on this analysis of my life from a biomimetic perspective, I'll try to remember that. Journey with me, and see how you fare.

I'll list each one of the Life's Principles and briefly summarize what I'm doing in this arena and what I could do better. Here goes!

Evolve to Survive

A big part of this principle is to replicate strategies that work; I do this through the integrated model for Alchemus Prime.[393] Personally, I use strategies for wellness, time management, and productivity that work for me. Integrating the unexpected is also an evolutionary strategy, and is something I do through improvisation techniques for myself and my clients, but sometimes it's challenging, because I'm a planner too. Reshuffling information happens through genetics in nature; I brainstorm and re-evaluate what I've done before in terms of products, services and facilitation, and then tailor based on what's needed. In a sense, I'm always creating too, if not procreating, ha! I think I could do better here by being more welcoming of the unexpected, and more nimble in the ways I integrate it. Having more time for play and fun would help me. Translation: I need to get back into dancing. It's been more than six months.

Adapt to Changing Conditions

Two of the strategies for adapting to change are to embrace diversity, and maintain integrity. I love diversity in all its aspects: ideas, ethnicities, ways of knowing, and more; without it, I feel unhappy. I grew up with 31 nationalities in my high school, and I'm sure that has a lot to do with it. Integrity is vital to me, and I practice self-renewal through meditation, personal retreats, and my poetry. Adapting to change is something I've made my career as a behavior change specialist, but it's still difficult on a personal level. The only solution is to accept and embrace change as the norm. I could always improve in my practice; it's a lifetime of learning to live this way. Nature's guidance is beautiful: plants and animals are rooted and serene in times of change, they never complain or worry, and they can't run away. They simply go with the flow.

Be Locally Attuned and Responsive

I'm good at being locally attuned in the sense that I walk everywhere and cultivate friendships with people in my neighborhood. I support local organizations and communities, and communicate frequently with those around me. The strategies of cultivating cooperative relationships, such as using feedback loops, and leveraging readily available resources are pretty familiar to me. I also welcome seasonal changes and cyclical processes, eating and doing what makes sense for the current season. To do better here, I need to use the previous principle to embrace change, as I had to do in my recent move to Southern California. Now, I'm in a process of learning my local neighborhood all over and starting anew.

Use Life-Friendly Chemistry

My biggest example of using simple, harmless chemistry is in the products I use for my body and dwelling. All are biodegradable and organic, affirming my values. I need to do better by bringing the same standard into all my materials for workshops and retreats. Often this is hard to control because we travel to different locations, but I could establish a practice of bringing my own cleaning products. A wonderful idea for a 2017 resolution for Alchemus Prime!

Be Resource Efficient

One of the big ways I'm being resource efficient is by eating low on the food chain. Plant-based foods take fewer resources such as land, water, and nutrients to grow, and are better for my health. I also ride a bicycle or take public transit whenever possible, and don't own a car to be more mindful of my resource use. I recycle, compost, and repurpose religiously whenever I can…but my biggest flaw is that I still fly. My dream is to one day have my own retreat and workshop venue, and have clients come to me using resource-efficient means. This would potentially reduce my travel by air to zero. Of course, I'm torn about this because traveling internationally affords so much intellectual growth and understanding of cultures, which may be difficult to give up entirely. Also, my family is spread out across the world, which makes traveling necessary periodically.

Integrate Development with Growth

The strategies at play here are working from the bottom-up, creating modular components, and self-organizing. I chose behavior change as a calling because it embraces these characteristics. I enjoy creating community and mobilizing bottom-up approaches because I firmly believe in scaling up individual behavior changes. I embraced modularity in the behavioral wellness strategy I designed for a company I consulted for. I want to do more in this vein, and mobilize masses of people to take back their health through plant-based eating and meditation.

My biggest take away from this introspection is that my learning is ongoing, and will take a lifetime. Onward!

Reflection:

Pick one biomimicry principle and describe how you're doing with respect to it. What might you improve?

Example from Margaret: Evolve to Survive — When I visit California in the winter, I drink fresh ginger tea to stay warm, because in Fiji, "winter" is about 75F. I need to be more mindful to wear warm clothing or layers, because I'm used to wearing just one layer in Fiji.

Three Biomimicry Principles for Engagement

I'm always exploring insights from biomimicry for social innovation.[394] My last post on biomimicry connected its principles to those of the Vipassana meditation technique.[395]

As I reflect on the work situations I've had the honor of participating in, from educational settings to local and international governments to startups, some lessons pop out that are in complete alignment with what biomimicry teaches us, and what I learned in my Biomimicry for Social Innovation Immersion course.[396] I'll summarize my top 3:

1. **Feedback loops:** One of the most important lessons that biomimicry reinforced for me is the importance of effective communication. Tight feedback loops, or quick, real-time communication, helps us navigate times of stress and change, allowing teams to co-evolve and strengthen their bond while also improving their productivity.

2. **Relationships:** According to master biomimicry teacher Dayna Baumeister of Biomimicry 3.8, "Life always puts relationship before task." In the workplace, this translates into being attuned to the whole person and our relationship with that person, and addressing any challenges there first, before jumping to task. Once the relationship is harmonious, the task will be handled with much more ease and effectiveness, because team members will be well motivated and more engaged.

3. **Adaptation:** Change is the only constant in life and work, so adaptation is a key skill and a lesson we keep learning. As conditions and resources shift, so does our approach to our processes, where we target our efforts, and how we execute our plans. The most important lenses to remember and apply here are the core DNA or core values of an organization, and the above two points: clear and timely communication, as well as attention to relationships, to ensure optimal adaptation. If an organization breaches its own core values and/or violates its relationships, the team, projects, and company fabric all start to suffer.

I continue to be endlessly inspired while learning from and applying biomimicry in my life and work. Share with me your key lessons learned from biomimicry for engagement in your workplace.

Reflection:

How might you apply biomimicry to engage your peers, students, or colleagues?

Example from Marilyn: I use a deck of biomimicry and leadership cards to engage my workshop participants to think creatively about their challenges as leaders.

Chapter 7 Notes

---

[346] Learn more about what biomimicry is: http://biomimicry.org/what-is-biomimicry/

[347] Read more about the three essential elements of biomimicry: https://biomimicry.net/the-buzz/resources/designlens-essential-elements/

[348] Definitions of leadership vary: http://www.businessnewsdaily.com/3647-leadership-definition.html

[349] See previous essay in this chapter.

[350] Read more about this innovative immersion workshop: https://biomimicry.net/what-we-do/professional-training/immersion-workshops/social-innovation-workshop/

[351] Learn more about this beautiful center: https://oaec.org/

[352] Learn more about biomimicry for social innovation: https://bio-sis.net/

[353] Life's Principles are a set of principles for how to learn from nature, using the discipline of biomimicry. Learn more here: https://biomimicry.net/the-buzz/resources/designlens-lifes-principles/

[354] Learn more about Google's supplier diversity program: http://www.google.com/diversity/suppliers/

[355] See Chapter 11 for more about the Alchemus Prime and our science-based Diamond Model.

[356] See the previous essay, entitled, "Pioneering a Harmonious Future: Biomimicry for Social Innovation."

[357] Research suggests that when we are stressed, we want to connect more with people: http://greatergood.berkeley.edu/article/item/how_to_transform_stress_courage_connection?

utm_content=buffercb419&utm_medium=social&utm_source=facebook.com&utm_campaign=buffer

[358] Read more about the tend and befriend tactic here: http://www.findem.com.au/resources/tendandbefriend.pdf

[359] See Note #292.

[360] See Note #288.

[361] Learn more about this amazing conference: http://conference.bioneers.org/

[362] Read about Toby Herzlich here: https://bio-sis.net/network/

[363] Learn more about acid rain: https://www.epa.gov/acidrain/what-acid-rain

[364] See the context in which we use this note: http://www.alchemusprime.com/services/

[365] See previous essays in this chapter for how we came to embrace and apply biomimicry.

[366] See the forum containing the translation here: http://somathread.ning.com/groups/taoism/forum/tao-te-ching-chapter-78-water

367 See the article here: https://www.brainpickings.org/2013/05/29/like-water-bruce-lee-artist-of-life/

368 Ibid.

369 More on true self in Chapter 11 and: http://www.alchemusprime.com/career-manifestation-program/

370 See essay entitled, "Biologically, We Are Nature," in this chapter.

371 See Further Reading section at the end of this book, and our website for more about Earth Champions: http://www.alchemusprime.com/our-books/

372 See the business-oriented translation here: http://www.wussu.com/laotzu/laotzu08.html

373 See Chapter 11, especially the essay entitled "In Search of the True Self: Alchemus Prime" for a discussion about how we incorporated the yin-yang concept into our corporate identity.

374 See Chapter 11 for the essay entitled "Science: Vulnerability is Beautiful."

375 Learn more about Michelle Oyen here: http://www.eng.cam.ac.uk/profiles/mlo29

376 Read the article summarizing her work: https://www.citylab.com/life/2016/09/a-skyscraper-made-of-bones-how-biomimicry-could-shape-the-cities-of-the-future/497969/

377 Concrete is a contributor to climate change: https://www.marketplace.org/2014/10/21/sustainability/surprising-contributor-climate-change-concrete

378 See Note #311.

379 See Note #311.

380 See Note #311.

381 See Note #296.

382 See Note #285.

383 See Note #305.

384 Learn more about how the black community is highlighting police violence against black women: http://www.aapf.org/sayhername-the-african-american-policy-forum/

385 Read more about the violence at Standing Rock: http://www.salon.com/2016/10/23/dakota-pipeline-showdown-at-standing-rock-when-a-powerful-corporate-chief-is-resisted-by-defenders-of-native-american-ceremonial-grounds/ as well as the most recent ruling: http://www.businessinsider.com/dakota-access-pipeline-ruling-2017-6

386 Learn more about BioNurse: http://challenge.biomimicry.org/en/custom/gallery/view/2525

387 Read more about the cushion plant: https://www.sciencedaily.com/releases/2013/02/130218092545.htm

[388] Read more about the Okanagan Four Societies from Nina Simon's keynote: https://medium.com/bioneers/reclaiming-relationship-tradition-towards-a-future-that-works-for-all-46ddac51d6d5

[389] See my blog post on engagement http://www.integritusprime.com/engagement-programs-boost-csr/ and Chapters 8, 9, and 10 for more on how to engage and change behavior for the right reasons.

[390] Life's Principles are a set of principles for how to learn from nature, using the discipline of biomimicry. Learn more here: https://biomimicry.net/the-buzz/resources/designlens-lifes-principles/

[391] Read more about Barbara Marx Hubbard: https://www.greenhearttransforms.org/event/evolutionary-woman-a-tribute-to-barbara-marx-hubbard/

[392] Conscious evolution is about how evolution evolves…read more about it: http://barbaramarxhubbard.com/conscious-evolution-defined/

[393] See Chapter 11 for more about the Alchemus Prime and our science-based Diamond Model.

[394] See also in this chapter, the essay entitled, "Pioneering a Harmonious Future: Biomimicry for Social Innovation."

[395] See the essay entitled, "The Biomimicry of Vipassana" in Chapter 6.

[396] Read more about this innovative immersion workshop: https://biomimicry.net/what-we-do/professional-training/immersion-workshops/social-innovation-workshop/

Chapter 8: Designing For Creative Engagement

Cadavers Teach Empathy

I usually think about empathy in the context of design thinking[397] and non-violent communication.[398] In this fascinating article from The Atlantic,[399] I learned that typically medical students are taught to become emotionally detached from their patients as they practice dissections on cadavers.

Now reversing this process somewhat, the University of Oklahoma College of Medicine is holding "Donor Lunches" where medical students meet the families of the deceased persons whose bodies they will dissect as part of their education in order to build empathy in the students. Interestingly, the idea was inspired by a Taiwanese medical school's practice where families of the deceased and medical students engaged in Buddhist prayers together prior to the dissections.

This is a wonderful example of a social innovation adapted for cultural appropriateness by Professor Jerry Vannatta, who created the Donor Lunches at UO College of Medicine. One of the results of this program is a more respectful and meaningful set of nicknames to replace the insensitive ones medical students previously came up with for cadavers. Kudos to Vannatta and to the Taiwanese for coming up with medical practices that re-integrate previously lacking empathy[400] into the healing practice of medicine.

And just as well, because a 1990 study out of Stanford University[401] suggests that medical students experience feelings similar to post-traumatic stress disorder (PTSD) when they cut open cadavers; this emotional distress needs to be addressed. Allowing medical students to feel empathy, respect and care for the deceased and their families offers a win-win solution for everyone involved, and puts the human(e) back into the medical profession.

Reflection:

How do you practice empathy?

Example from Marilyn: It's very typical and easy for me to feel what another person or even animal is feeling, so I practice empathy by being aware of what I'm feeling and do my best to separate which feelings are mine, and which are not. Then I can be compassionate without getting involved in any way that compromises my strength.

We are the Tech We've Been Looking For

In 2015 I traveled to Urcuqui in Ecuador for the launch of the Innopolis Science and Technology Fair in Yachay City of Knowledge.[402] A wonderfully ambitious project, Yachay aims to be the next Silicon Valley.

As a behavioral scientist, I was a minority at this event. I mingled with badass biologists, like Dr. Barry Bruce from University of Tennessee Knoxville, who works on making electricity from photosynthesis, which he aptly calls 'growing electricity'. I also connected with innovation gurus, like Greg Horowitt, serial entrepreneur and venture capitalist who divides his time between Stanford and UC San Diego, and Jeremy Abbett, Creative Evangelist at Google in Hamburg.

I joined hundreds of Ecuadorian youth admiring the drone demonstration Jeremy led. Artificial intelligence devices, 3D printers, and all kinds of nifty gadgetry in the fab lab surrounded me. I also shook hands and chatted with the Vice President of Ecuador, Jorge Glas, and learned that he used to be Energy Minister, and is an electrical engineer by training.

My talk, on a panel focused on renewable energy technology transfer, was about the human components of technology transfer. I spoke about the need for sharing and nurturing human knowledge, skills, and process design to, as my former mentor the late Steve Schneider advocated,[403]

'…leap frog entirely the Victorian industrial revolution to a high tech, low carbon, and very efficient industrial future.'

My points were made more salient by the unexpectedly poor coordination of the Yachay event. There was state-of-the-art technology everywhere, but those who were working with simpler communication and transportation technologies, were stressed and under extreme time pressure due to circumstances beyond their control.

A year before, I was in Tena, Ecuador, participating in the IKIAM Amazon University's Academic Review Workshop. Also an admirable and ambitious effort, IKIAM aims to launch with learning-by-doing as its primary teaching and learning style across undergraduate and then graduate programs. While IKIAM's event organization was impeccable, there have been delays in the implementation of the working groups' recommendations to train incoming faculty in learning-by-doing methods.

During both visits, two somewhat opposite forces in this beautiful country struck me. First, the strong and truly admirable ambition to fundamentally transform education and innovation in Ecuador was clear and present. Second, the effect of what appears to be governmental hierarchy and bureaucracy that may interfere with fulfilling this ambition, was also crystal clear.

Ecuador is undertaking a mission that requires fundamental cultural change. Such change takes time, and government leaders are to be commended for taking bold steps forward.

For IKIAM, design thinking and place-based learning techniques need to be taught to government administrators as well as IKIAM faculty to build unity toward a smooth implementation of IKIAM's beautiful vision of learning-by-doing. For Yachay, I would recommend something similar: design thinking for team building, improved communication, and event management so that the human technology – knowledge, skills, and process management – are in sync with the physical technology for a smoother implementation of events such as the launch of Innopolis.

In my talk, I encouraged the inaugural students of Yachay University to look within and invest in learning the skills, knowledge, and techniques they need to work effectively with technology to move their amazing nation forward.

Earlier, in a discussion with my fellow presenters, I pointed out that technology is created, implemented, repaired, and improved by humans; Greg Horowitt responded, *"That has always been the case."*

Technology is only as effective as we are. Human ingenuity is our core treasure in the world of innovation – we are the tech we've been looking for.

Reflection:

How might you begin to creatively foster positive change in a bureaucratic, rigid, or closed culture?

Example from Marilyn: I would use short, fun processes like improvisation, brainstorming and prototyping to prove that change can be sparked quickly, and then scale the effort using workshops including every member of the group so that everyone speaks the same design thinking language and, over time, can collaborate well.

Design Thinking: Reframing for Success

A recent New York Times article[404] on the personal and professional applications of design thinking is after my own heart. One of the critical benefits of design thinking, as my Stanford colleague Bernie Roth explains in this article, is taking the step of reframing the problem. He states,[405]

"If you have tried something and it hasn't worked, then you're working on the wrong problem."

When I was co-teaching one of the workshops in the Research as Design[406] program, which a team of us co-founded to apply design thinking to help scholars innovate with their research processes, we encountered a first year graduate student who was having trouble coming up with research questions. As we probed his dilemma, however, we discovered that coming up with research questions wasn't the problem. The real issue was fear of being judged by his advisor. Feeling intimidated by brilliant professors at an institution like Stanford is not uncommon. Our young participant reframed his challenge and came up with solutions for how to overcome his fear of approaching his advisor, including practicing talking about his research ideas with friends and colleagues first, as a prototype.

Over the past seven years I've facilitated workshops in academia and for companies, nonprofits and interdisciplinary conferences, and I keep finding the same pattern: we are conditioned to think in certain ways, which narrows our solution pathways. Snapping out of our habitual ways of thinking allows us to reframe our challenges more accurately and then find ways to change our behavior to achieve our goals.[407]

As the author of the New York times article points out, she was able to lose 25 pounds when she reframed her "I have to lose weight" challenge as "I want to spend time with my friends socially." The latter reframe, and self-empathy, allowed her to motivate herself to go out more, eat better, and ultimately, lose weight.

As we embark on our new year's resolutions and organizational targets year after year, it's important to consider if we are focusing on the correct problem, and design thinking can help.

Reflection:

What is a challenge you're facing right now? How might you reframe it?

Example from Marilyn: I was recently thinking about job options that would help me make a larger positive impact on climate change reduction. I reframed this as a challenge that was not about jobs per se, but more about multiple pathways to leveraging my skills. We cannot be forced into a box: we are destined to spread our wings and do many things. This book is a way of sharing my ideas for a better world by empowering ourselves to make changes starting today!

Design Thinking at Scale

IBM is applying design thinking to change the way its 400,000 employees work.[408]

Design thinking, or human-centered design, is a dynamic and flexible process for quick problem-solving from the perspective of the user, or the recipient of the solution. How might IBM use it to change its operations?

Well, IBM has adapted design thinking for its own uses, coming up with a process called The Loop.[409] The Loop consists of an infinity symbol, which contains three green dots and a yellow dot, all four forming a square. The loop represents infinite, or what they call "restless" innovation, while the green dots represent multidisciplinary teams, and the yellow dot represents the user. The three principles are to constantly innovate, focus on the user, and use multiple perspectives through their teams.

How does this giant company build in agility? After all, it's a complex working environment and design thinking is designed to craft clarity, albeit typically for small groups. The mantra IBM has adopted is: everything is a prototype.

It's difficult not to love this approach. The rule for prototyping in design thinking is to fail early and often, before many resources become invested in a project or idea. To view everything as a prototype, to me, encompasses several wonderful principles, including the following:

1. **Detachment:** if it's a prototype, you won't be too attached to it as a final product, which means you'll be open to other ideas, AND you'll constantly be thinking of ways to improve on them.
2. **Mindfulness:** chances are you'll stay in the moment, and cultivate a process-oriented mindset, focusing on the prototype and its current evolution instead of jumping to conclusions or implementation.
3. **Iteration:** since you're prototyping, you can follow a different angle or start over with little grief if something fails (i.e. doesn't meet the criteria for user satisfaction).

Of course, it's also necessary to implement, finish or deliver a polished product or service. However, once that's done, the product can *still* be viewed as a prototype with an eye to improvement. A work environment where everyone is always learning builds in engagement, constructive feedback, and productivity, and leverages unpredictability, while banishing negative judgment, drudgery and routine. This way of working also flattens company hierarchy and with it, removes barriers to communication and cooperation.

IBM is following through on its commitment to mainstreaming design thinking: the ratio of designers to coders has gone from 1:80 to 1:20, and the target is 1:15. Also, the company has put 10,000 employees through its design bootcamp to date – while this is only 2.5% of employees, 100 products have also emerged from their design thinking processes.[410] It will be interesting to see how IBM continues this journey of integrating design thinking into its entire operations.

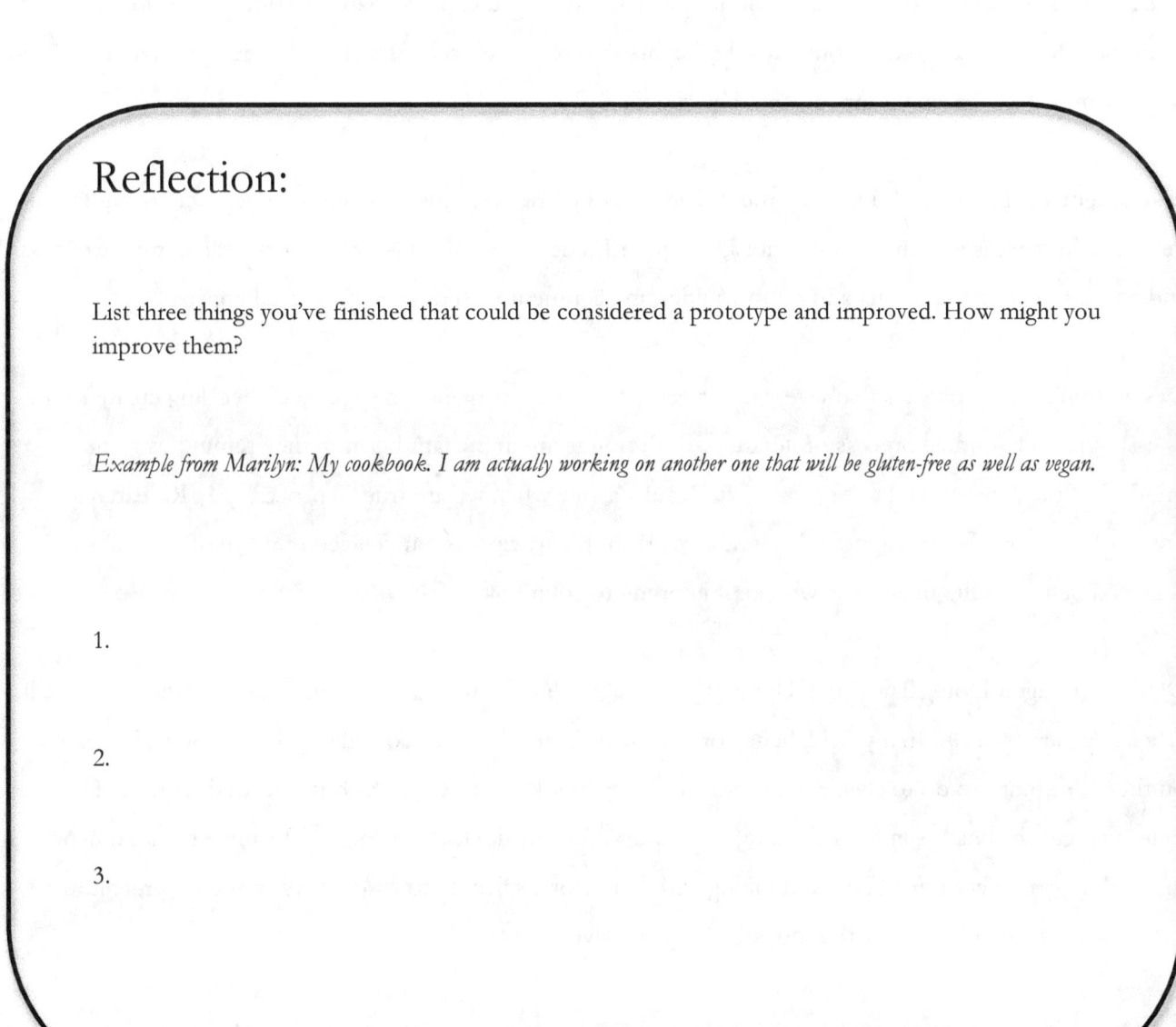

Reflection:

List three things you've finished that could be considered a prototype and improved. How might you improve them?

Example from Marilyn: My cookbook. I am actually working on another one that will be gluten-free as well as vegan.

1.

2.

3.

Design Thinking: Double Diamond

I recently read an article[411] about revamping British Design Council's[412] double diamond[413] that really resonated with me. The double diamond is a framework for how to apply a design process from the discovery phase of a challenge to the delivery of the solution. This article is well-written, with simple language and useful examples. Thank you to Master's student Dan Nessler for doing such a stellar job.[414]

The article is so good that it's best to read it in its entirety,[415] and explore Dan's version of the double diamond. However, I will tease out a few highlights that really gel with Alchemus Prime's approach, and add some commentary based on my research and practice.

Fundamentally, I agree with Dan that the design thinking process initially requires a deep understanding of people (I frame this as human motivation), although I tend to expand this phase to include human wellness and environmental constraints and opportunities, in alignment with our science-based approach.

Design thinking is a process that leverages divergence and convergence in sequence, revealing creativity as essentially a continual process of iteration. Both points are important because they remind us to be mindful of the process, and to be loyal to it. Results arrive when we are true to process. My Research as Design[416] work made this abundantly clear, as workshop participants who stayed mindful of process achieved better results than those who kept jumping to solutions.

Dan's two stages, Doing The Right Thing (what I call being effective), and Doing Things Right (what I call efficiency), are spot on. In my field, behavior change, it is important to consider the vast body of research on the right things to do to change behavior, so design thinking fits better into the second stage, of HOW to implement the change in ways that are motivating.[417] In my doctoral research[418] I came up with a hybrid methodology that combined design thinking with behavioral sciences to study ways to reduce greenhouse gas emissions, and come up with innovative ways to save energy.

Lastly, I love that Dan openly stated that his revamped version of the double diamond is just that, a version, and that each problem-solving process is different and will require tweaks and adaptations. At Alchemus Prime, our model is built to generate tailored processes and tools so that we can focus with fresh lenses and combine tools in new ways for each unique challenge our clients bring us. Iteration is our friend!

Reflection:

What is a task you do often in which you could be more mindful of the process to get better results?

Example from Marilyn: Walking – sometimes I get fixated on the destination, but when I'm mindful of the process and the journey, I have a much more relaxing and inspiring walk.

Knowing Your Fish: Data-Driven Design

At Alchemus Prime, my business partner often gives an example to guide our clients with business strategy. We call it the fish story. Basically, you could buy and use all the fancy equipment, and gear you want, but at the end of the day, the fish only cares about bait. So, are we offering what the customer wants?

In design thinking, we call this human-centeredness,[419] which is an over-arching aspect of the process. Similarly, in lean startup methodology,[420] almost every step is focused on the customers and what should be produced for them. The upshot of both approaches, which overlap significantly,[421] is that designing successful solutions depends on continuous learning from, and adapting to, the data gathered from the user.

The foundation for our ability to serve clients well is everything we learn from clients about their past experiences, present situation and challenges, and aspirations for the future. Our data collection then drives our service offerings. In essence, the client is helping us design the solutions they need, through their intimate knowledge, wisdom and context, which we simply draw out and reimagine as solutions with them.

For example, we developed our Career Manifestation Program[422] after informal and formal conversations and interviews with high achievers in our networks, which led us to realize their need for a career reinvention retreat with high consciousness and integrity. Our conversations and interviews yielded rich data, which we fed into our Alchemus Prime Diamond Model.[423] The model integrates design thinking with behavioral science, biomimicry, and meditation. We developed our career manifestation tools, and refined them as we facilitated more and more retreats. Over time, we implemented data-driven changes to our program, including in the duration, content and format, and post-retreat follow up.

We found that our "fish" consist of individuals, teams, and organizations in various stages of development. They all desire a deeper harmony between their financial, professional, personal, social, environmental, and recreational goals. Our clients are self-aware and highly accomplished achievers who are ready to rise to the next level of success, which they define as offering the best of one's skills, resources, assets, and networks to clients in ways that achieve one's own goals, and generate benefits for one's family, community, and the planet. These leaders want to be true to themselves, and live from a core of inner and outer alignment with life.

We are nurturing a paradigm of integrated methodologies for win-win partnerships and solutions that keep each participant engaged, fulfilled, and abundant.

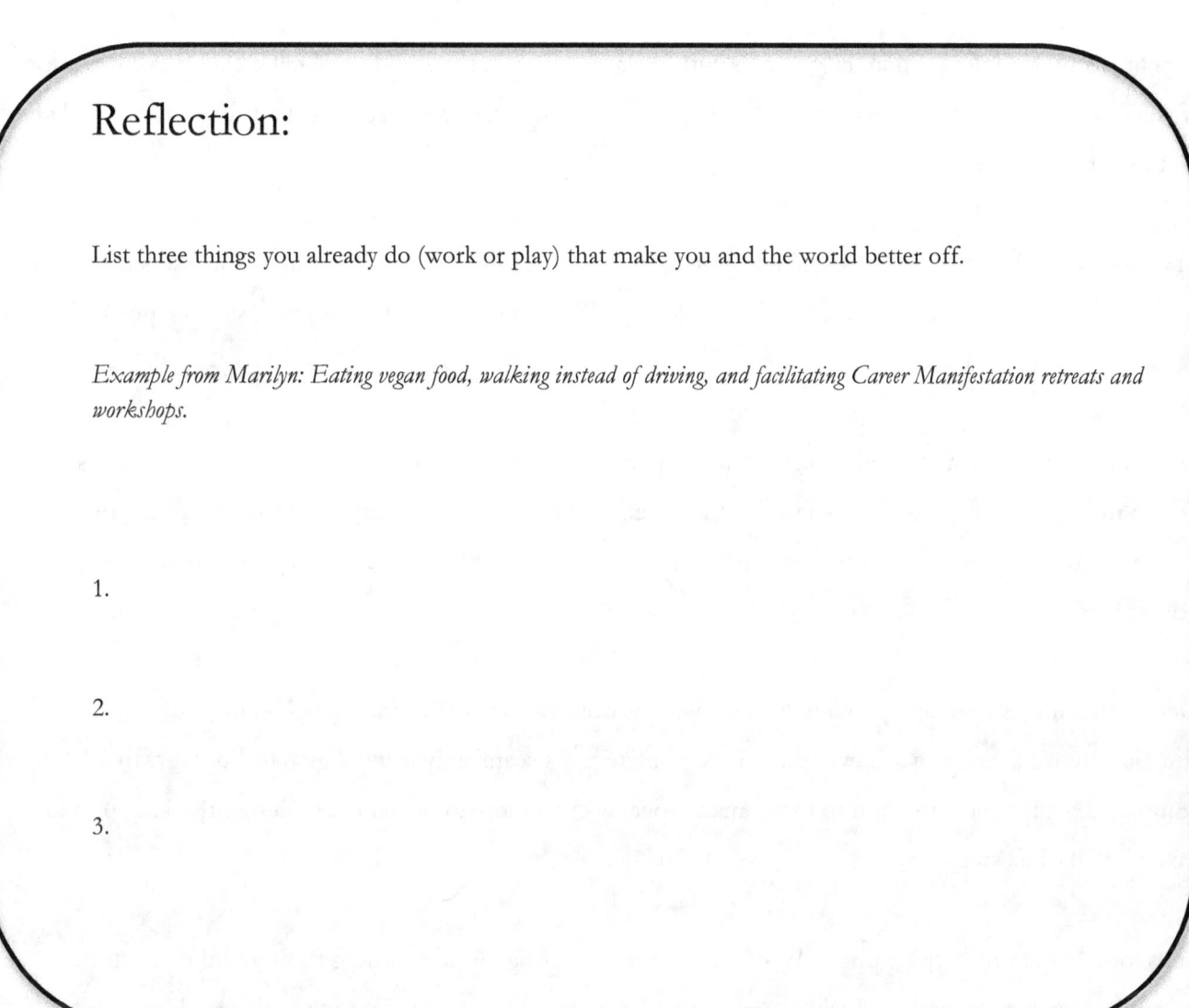

Reflection:

List three things you already do (work or play) that make you and the world better off.

Example from Marilyn: Eating vegan food, walking instead of driving, and facilitating Career Manifestation retreats and workshops.

1.

2.

3.

Design Thinking for Engagement

Recently, I melded design thinking process with behavioral science methods to facilitate a breakout session at Acterra's Spring Green Team Forum.[424] My goal was to re-invigorate green team representatives, and give them fresh tools.

In a busy workplace with many seemingly competing goals, it can be difficult to rally co-workers around sustainability. We received positive feedback from the 32 participants, and once again, saw the power of design thinking at work.

Design thinking is being touted as a galvanizing methodology for companies. In digital learning circles,[425] design thinking is being applied as a tool for innovation. Leveraging the ideation or brainstorming phase of the design thinking process, employees can come up with divergent perspectives and use them to create novel solutions.

Design thinking is also being used to achieve what is called a state of Digital Flow.[426] This refers to situations when a user can achieve goals and complete tasks seamlessly using digital technology. In a complex digital world, attention to the human aspect is even more important, and design thinking provides that with all of its steps.

Let's look at the prototyping phase. Wisely, user interface designers are turning to more human-centric prototypes such as scenario modeling[427] and role play, which I have co-taught in the Research as Design[428] program at Stanford University and in climate and sustainability action planning sessions for state and international governments. Behavioral prototypes provide insight into what could go wrong, how to anticipate and satisfy questions, and how to establish and build trust. These lessons also apply to staff retention as well as to customer acquisition and cultivating long-term relationships.

Broadening our scope from the ubiquitous digital era and recognizing its human foundations, it's important to embed design thinking into the culture of a company.[429] We can do this in various ways, including starting with a particular step, and then slowly incorporating more phases of the design thinking process into daily or weekly operations. One of our former clients chose to embed ideation into their Monday planning meetings, and found it to be a very time-efficient and fun way to start the week.

Reflection:

Describe how you use brainstorming or prototyping in your work or personal life to engage yourself or others. If you aren't, how might you start?

Example from Marilyn: I use brainstorming and prototyping to help clients get unstuck, generate many solution ideas, and build out some of them. They usually leave with more confidence and a set of actionable solutions.

Chapter 8 Notes

---

[397] Design thinking is a process for generating creativity. Read more on our website: http://www.alchemusprime.com/design-thinking/

[398] See the two essays in Chapter 5 about my weekend in prison, entitled "Seeing Beyond Orange…"

[399] Read the full article here: https://www.theatlantic.com/health/archive/2015/07/cadaver-dissection-empathy-medical-school/398429/?utm_source=SFFB

[400] Ibid.

[401] Learn more about this study: http://onlinelibrary.wiley.com/doi/10.1002/ca.980030308/abstract

[402] Read more about the launch of this event: http://www.fablabconnect.com/ecuadorian-vice-president-opens-the-innovation-event-of-the-year-innopolis/

[403] See the late Steve Schneider's paper here: http://www.accc.gv.at/pdf/schnei.pdf

[404] Read the article here: https://well.blogs.nytimes.com/2016/01/04/design-thinking-for-a-better-you/?smid=tw-share&_r=1

[405] Ibid.

[406] Read more about the Research As Design program here: https://researchasdesign.com/

[407] See Chapter 9 for more on behavior change.

[408] Read more about IBM's approach here: https://www.wired.com/2016/01/ibms-got-a-plan-to-bring-design-thinking-to-big-business/

[409] Read more about The Loop: https://www.ibm.com/design/thinking/

[410] See Note #343.

[411] Read more about design thinking and the double diamond here: https://medium.com/digital-experience-design/how-to-apply-a-design-thinking-hcd-ux-or-any-creative-process-from-scratch-b8786efbf812#.mig79dvli

[412] More on the British Design Council: http://www.designcouncil.org.uk/

[413] The British Design Council's Double Diamond framework for design thinking: http://static1.squarespace.com/static/55fa0341e4b06660c65bd4f0/t/5642c682e4b0b633d4fcc1fd/1447216776499/

[414] Read more about Dan Nessler: http://www.dannessler.com/

[415] See Note #346.

[416] See Note #341.

[417] Read more about how to apply design thinking in the "how to implement" stage of a project: https://researchasdesign.com/2012/01/01/applying-design-thinking-to-interdisciplinary-research-energy-and-behavior-2/

[418] Learn more about how my doctoral research integrated behavioral sciences with design thinking: https://earth.stanford.edu/events/e-iper-dissertation-defense-marilyn-cornelius

[419] Learn more about user-centeredness: https://www.usertesting.com/blog/2015/07/09/how-ideo-uses-customer-insights-to-design-innovative-products-users-love/

[420] Learn more about lean startup methodology: http://theleanstartup.com/principles

[421] Learn more about the overlaps between lean startup methodology and design thinking from the Graduate School of Business at Stanford: https://www.gsb.stanford.edu/insights/lean-startup-design-thinking-getting-best-out-both

[422] Learn more about our Career Manifestation Program: http://www.alchemusprime.com/career-manifestation-program/

[423] See Chapter 11 for more about the Alchemus Prime and our science-based Diamond Model.

[424] Learn more about Acterra's work with green teams: https://www.acterra.org/greenteamnetwork/

[425] Learn more about digital learning circles: http://www.clomedia.com/2017/03/02/7-ways-boost-employee-engagement-digital-era/

[426] Read more about digital flow: https://www.forbes.com/sites/alfresco/2017/03/10/capital-one-embraces-design-thinking/#7d70850130e7

[427] Read about an example of scenario modeling: https://www.forbes.com/sites/alfresco/2017/03/10/capital-one-embraces-design-thinking/#7d70850130e7

[428] See Note #341.

[429] Embedding design thinking into the culture of a company is important. Read more: https://blog.shrm.org/blog/why-design-thinking-and-why-now

Chapter 9: Behavior Change - The Ultimate Solution

On Motivation: Love and Real-Time Feedback

Do you remember the last time you were in love? Or perhaps the last time you were with a loved one – a beloved family member or close friend. Pause for a moment and consider how you felt and what you were doing. Chances are, you were enjoying yourself, whatever you were doing.

This is because we humans like to do things that we enjoy. Sounds pretty simple, right? That's because it is!

Doing an action because we enjoy it is an example of intrinsic motivation.[430] Extrinsic motivation, on the other hand, is when we do something for a reward. Think about it this way – you're washing a huge pile of dishes for your friend because she's rushing off to a very important job interview and you want to help out so her guests don't come home to a messy kitchen later tonight. She slaps a five-dollar bill on the counter and says, *"Thanks so much!"* How do you feel? Your priceless kind deed has now been valued at five dollars.

We're wired to connect and empathize[431] with others and feel like we belong to a group or community. When we do things in this context, it's like play, i.e. very motivating. When play is turned into work, we become less motivated.[432]

Companies like Google have taken this insight and applied it to make their employees feel more playful and creative at work. I visited their San Bruno YouTube office and noticed the funky and colorful décor, as well as the variety of spacious indoor and outdoor spaces, and lots of dogs! IDEO and the d.school at Stanford are designed in similar ways to inspire creativity and mirth.[433]

As Dan Pink explains in his book, Drive,[434] we are also motivated by a sense of purpose, and the desire to create something meaningful. These motivations are salient, or strong, and can be lasting and most satisfying compared to external rewards like awards or raises. Of course, it's nice to be compensated fairly and generously, but that's not what drives us.

Once we are carrying out the actions we want to because we care about making a difference or helping a loved one, behavioral tools can be used to keep us motivated along the way. For instance, setting goals that are specific, measurable, attainable, relevant, and time-bound (in other words SMART)[435] and can be very motivating. And, if you've ever run on a treadmill or worn your Fitbit or other wearable device while walking or running, you know that receiving instant, real-time feedback from the display can be crucial to keep you going till you meet your goal. Sometimes, I get so motivated, I exceed my goal.

Why not leverage this knowledge to motivate ourselves to be the best we can be, and make meaningful contributions to the world? Well, we are! As we move toward a life with more balance, more creativity, and making contributions to nurturing life through the human-nature innovation engine I call Alchemus Prime, we can leverage these powerful motivations and behavioral tools to help be the best we can be, and to achieve what previously may have seemed impossible. Impossible is the new goal. Let's go!

Reflection:

List three tasks you find fun and motivating, and three that you do not. How might you leverage the first set to accomplish the second set of tasks?

Example from Marilyn: I love watching movies, but hate doing taxes. I often reward myself for finishing my taxes with an outing to a movie I know I'll love.

Fun:

1.

2.

3.

Not so much:

1.

2.

3.

Connecting Behavior Change and Biodiversity

Time published an article[436] about a study[437] on climate change and when I read it, I was dumbstruck by two points the article made:

1. Restoring nature, biodiversity in particular, is important to address climate change (This is hardly news to anyone, because biodiversity brings more resilience and capacity to cope with climate disturbances).

2. As an alternative to changing human behavior, we should try to conserve biodiversity (This falsely implies that these two activities mutually exclusive).

Okay, I can't help but be a little flabbergasted...First, please don't tell me science has become so compartmentalized that we are now talking about climate change and nature as separate concepts. Climate change is an aspect of nature, right? It's not an alien phenomenon, although we humans have accelerated planetary warming unnaturally, which is why we are facing extreme events and other dangerous impacts.

To be fair, the article mentions the connections between nature and climate change in terms of policy to reverse deforestation. The Intergovernmental Panel on Climate Change (IPCC) reports[438] concur that halting and reversing deforestation is a critically important step in fighting climate change. Okay, but how? I'll come back to that.

Second, why is restoration of biodiversity considered an approach separate from human behavior change? Why are we experiencing a mass extinction in the first place? Is deforestation not human behavior? Human demand for more meat, dairy, seafood, cars, cell phones, and so many other products is leading to the rapid destruction of forests, grasslands, oceans, and other ecosystems. Behavior change cannot be restricted to mean only efficiency actions such as turning off lights. There are many other behaviors we need to look at holistically.

Let me explicitly build the bridge between human behavior and loss of biodiversity: our daily actions are causing the destruction of ecosystems, which contain the precious biodiversity that can offer resilience and adaptation in a rapidly warming world. As a study based on integrated land use system modeling indicates,[439] our behavior change to a vegan lifestyle would free up grasslands and pasturelands, which can then be restored to native forest to sequester more carbon than what humans have added to the atmosphere since the industrial era began. That, my friends, is the inextricable connection between our behavior, and the biodiversity required for resilient ecosystems that can cope with climate-induced extreme events. Veganism is an important component of the how.

I'll leave you with this science-based point: biologically, we are nature,[440] and it's time to start acting that way. The future of the planet lies in the power of our daily actions, and if each of us changes our behavior to reverse deforestation and save biodiversity, we can make a difference faster than we expect. As the Center for Biological Diversity has said, take extinction off your plate.[441]

Reflection:

Draw or write the connections between one product you use and biodiversity or wildlife.

Example from Marilyn: I drink coffee rarely. When coffee is grown in native forests e.g. in Costa Rica, this can result in wildlife being forced out of its natural home, and the loss of precious species.

Ending Procrastination

I recently read a blog post[442] by fellow behavioral scientist Frank Martela[443] with great interest because it resonates with my behavioral science training at Stanford, which I'm humbled to say, included research by Al Bandura[444] and Mark Lepper,[445] two of the godfathers of social psychology.

In the post, Martela writes about barrier analysis and barrier removal. In behavioral science speak, a barrier is anything that stops you from doing what you want. For example, not being able to find your gym shoes might prevent you from going to the gym the next day. To overcome the barrier, you might find your shoes the day before and place them next to your front door.

Similarly, research by the famous Brian Wansink[446] and others[447] suggests that small and seemingly insignificant factors can radically change our behavior, in particular with relation to our eating habits. One of the more well known changes we can make is simply stocking our kitchen with smaller plates and bowls: we automatically eat less!

Similarly, we can hack our current negative habits by reframing the first 20 seconds of the behavior. As Shawn Anchor discusses in his book, The Happiness Advantage,[448] removing the batteries from his remote and placing them in another room, and placing books he wanted to read on the coffee table enabled him to stop watching so much TV and start reading more. From a behavioral standpoint, we are overcoming barriers to the desired behavior (books are too far away, so we place them closer) and creating barriers for the undesired behavior (remote is too readily accessible, so we place it far away), in order to make the desired behavior (read more than watch TV) easier to accomplish.

So, the next time you're making a new year's resolution or simply wanting to start a new habit or stop procrastinating, take into account the steps that will make your desired behavior easier to achieve, and put those 20-second rules into action!

Reflection:

What is one thing you can do today that will help you with the first 20 seconds of something you've been meaning to do for a while? (Tip: after you write about it, do it).

Example from Marilyn: I need to start lifting weights again, and the way to make this task easier is to place the weights where I can see them and connect them to my workout – probably next to my yoga mat.

Engagement Approaches That Work

Engagement means being interested, active, and feeling good about our work in the workplace. It's important to understand the immensity of the engagement challenge before us. A staggering 69% of managers surveyed[449] feel uncomfortable communicating with their employees, and 37% of business leaders are uncomfortable giving direct feedback.

Pause for a moment. Communicating and giving clear feedback are essential to working effectively. Without these skills, we might as well be sailing without a compass.

Let's look at some approaches that work. The same survey, by Interact,[450] and reproduced in the Harvard Business Review,[451] lists strategies for how to effectively give feedback:

1. Be direct and kind
2. Listen
3. Don't make it personal
4. Be present
5. Inspire greatness

Easier said than done, and it takes practice.

Besides effectively communicating and giving feedback, other skill sets are crucial to keeping our employees inspired and productive. The very gifted and best-selling author Seth Godin,[452] in his excellent blog, "Let's Stop Calling Them Soft Skills,"[453] presents an index of what he calls real skills, in five categories. This list is exceptionally perceptive because it captures traits that distinguish a leader from others with a similar resume who might lack these skills. I've summarized the list below in my own words:

- **Self-Control** – being able to do what it takes to persist with achieving goals and priorities, through adaptive capacity, integrity, resilience in the face of failure, compassion, a collaborative approach, willingness to learn from feedback, enthusiasm, ability to manage stress and be flexible with change, confidence and passion, and sound ethics.

- **Productivity** – the ability to produce results and keep moving forward through sound decision making, diligence, delegation, goal setting, attention to detail, crisis management, creative problem-solving, planning, research, time management, facilitation of discussions, and testing, learning, and adapting.

- **Wisdom** – the knowledge we gain from experience, not textbooks, and apply through our artistic sense and creativity, instincts for mediation and dealing with difficult people and situations, showing empathy across diverse settings, supervising and mentoring, and exercising sound emotional and social intelligence.

- **Perception** – the art of seeing things accurately for what they are and could be, through design thinking, esthetic and fashion sense, evaluating people and situations intuitively, and creating mind maps and strategic plans.

- **Influence** – the ability to move others to take action, through delivering clear feedback and constructive criticism, appropriate body language, written and verbal language skills, ability to speak, reframe, sell, tell stories, and present ideas in compelling ways, and charisma.

Studying these qualities affirms our thinking and approach to nurturing leaders and inspiring engagement: it begins with the true self.[454] Charisma and confidence don't come from outside us; we can foster them from within by facing fears and diligently learning and practicing the skills and traits we want to hone.

These "real skills" remind me of two resources I loved during graduate school: *Crucial Conversations*[455], which is about the importance of having difficult conversations in order to be a leader. And, as a behavioral scientist, I also loved another book by the same authors, called *Influencer.*[456]

One of my greatest experiential lessons through facilitating workshops is that my confidence, passion and enthusiasm can be contagious. Another is that persistence with those who are slow to open up, or afraid to fail, or shy, pays off in brilliant solution ideas once they overcome the fear of failure or discomfort. Once a person feels cared for, and has permission to play, fail, and be silly, all things become possible.
I chose engagement and behavior change because I love helping people and seeing the sparks of inspiration in their eyes – my calling is a gift to me, and I hope, to all I serve.

Reflection:

How do you give feedback? How might you improve your style?

Example from Marilyn: I am very direct and encouraging at the same time. I can improve by being mindful of remembering to start with positive feedback and ending with positive feedback, with constructive criticism in the middle.

Mindfulness, Values, and Behavior Change

At Alchemus Prime, we are very mindful of our core values. We want to build relationships with partners and customers, and together nurture leaders and organizations in ways that ultimately benefit the planet. Being mindful allows us to observe situations from a distance. This spaciousness allows us to detach, make decisions and set healthy boundaries based on values.

For example, as a graduate student studying the behaviors that negatively impact climate change, I came across some (now famous) United Nations research[457] on the impacts of food on our climate. In particular, I discovered that animal agriculture,[458] especially meat and dairy products, contribute more to climate change globally than anything else. Around the same time, after one conversation about the impacts of the dairy industry on calves (they are separated from their mothers but kept just close enough to keep the cow producing milk – torture, anyone?), I went vegan.

Two important core values were activated for me:
1. I am on a personal and professional mission to reduce and reverse climate change.
2. I support animal freedom and wellness in every way I can.

Recently, I came across another powerful example of how values can change behavior. Harvard psychologist Susan David demonstrates how values can change our behavior instantly. She explains that while we once might have succumbed to chocolate cake in the past; linking that unhealthy eating behavior to the orphaning of our children can immediately steer us in the direction of better dietary choices. Watch the six-minute video – it's powerful![459]

These three examples of how values drive behavior bridge personal and professional choices, illustrating the importance of mindfully using our core values as an internal compass that guides our outer actions in the world.

On a less serious, and more delicious note, *any* chocolate cake I eat is climate and animal friendly. Share your stories about how a core value led to a change in your habits, chocolate cake and beyond!

Reflection:

Describe one example where your values and actions are aligned, and one example where they are not. How does each one feel?

Example from Marilyn: I am vegan and this fits my core values – and I am at peace about it. I still fly and this has a high carbon footprint, so this bothers me a lot.

Aligned:

Misaligned:

Chapter 9 Notes

———————————————

[430] Read more about intrinsic motivation: http://study.com/academy/lesson/intrinsic-motivation-in-psychology-definition-examples-factors.html

[431] Check out the book *Empathic Civilization* to learn just how much we are wired for empathy: http://www.empathiccivilization.com/

[432] Read more about the consequences of turning play into work: http://psycnet.apa.org/psycinfo/1975-21035-001

[433] Check out the IDEO and d.school office spaces on their websites: https://www.ideo.com/ https://dschool.stanford.edu/

[434] Read more about what motivates us in the book *Drive*: http://www.danpink.com/books/drive/

[435] Learn more about SMART goals: http://hrweb.mit.edu/performance-development/goal-setting-developmental-planning/smart-goals

[436] Read the Time article here: http://time.com/4070683/nature-climate-change/?xid=fbshare

[437] Read the scientific study published in the journal *Nature*, here: http://www.nature.com/articles/nature15374.epdf?referrer_access_token=yWJ957J9aqp9gSNjodpQjdRgN0jAjWel9jnR3ZoTv0MECCCA2xlBG3fiA2jnrlp2IGLF31Bw4yA1Uc1vRZdF2-C80kMWX14aAYCjyBLOKjfSqn_1480yEoYBwX-boYVMsMZT_wTw2jsWzurTgOy8myBKUALNA8ogvbg2Vzf9_Kan2walkIWb4_cazwC5hdT2&tracking_referrer=time.com

[438] Read about the reports here: https://www.cgdev.org/blog/tropical-forests-offer-24–30-percent-potential-climate-mitigation

[439] Read more about the study's findings here: https://agu.confex.com/agu/fm15/meetingapp.cgi/Paper/67429

[440] See Chapter 7 for more on how and why humans are nature.

[441] Learn more about this campaign: http://www.takeextinctionoffyourplate.com/

[442] Read the full post here: http://www.fulfillmentdaily.com/20-second-rule-ending-procrastination/?utm_content=buffer201cb&utm_medium=social&utm_source=facebook.com&utm_campaign=buffer

[443] Read about Frank Martela: http://www.fulfillmentdaily.com/author/frank/

[444] Read about Al Bandura: http://stanford.edu/dept/psychology/bandura/bandura-bio-pajares/Albert%20_Bandura%20_Biographical_Sketch.html

[445] Research by Mark Lepper: https://stanford.academia.edu/MarkLepper

[446] Read about Brian Wansink's research: https://dyson.cornell.edu/people/brian-wansink

[447] More research on food cues: http://jamanetwork.com/journals/jamapsychiatry/article-abstract/492411

[448] Check out *The Happiness Advantage*: https://www.amazon.com/Happiness-Advantage-Principles-Psychology-Performance-ebook/dp/B005L193RO/

[449] Read more about these findings here: http://interactauthentically.com/new-interact-report-many-leaders-shrink-from-straight-talk-with-employees/

[450] Ibid.

[451] See the Harvard Business Review article: https://hbr.org/2016/03/two-thirds-of-managers-are-uncomfortable-communicating-with-employees

[452] Read more about Seth Godin: http://www.sethgodin.com/sg/bio.asp

[453] Read the full blog here: https://itsyourturnblog.com/lets-stop-calling-them-soft-skills-9cc27ec09ecb?gi=c7ad7730932c

[454] See Chapter 11 for more on the true self.

[455] Learn more about this book: https://www.vitalsmarts.com/resource/crucial-conversations-book/

[456] Learn more about *Influencer*: https://www.vitalsmarts.com/resource/influencer-book/

[457] Read more about how food impacts climate change in Chapter 3 and through this research: http://www.europarl.europa.eu/climatechange/doc/FAO%20report%20executive%20summary.pdf

[458] Ibid.

[459] Watch the video here: https://www.youtube.com/watch?v=0_6hu6JLH98

Chapter 10: How To Act On Climate Change

Remembering Steve Schneider, Climate Warrior

It's been about seven years since our planet lost its most intrepid climate warrior. I want to start this chapter with a remembrance of Steve because he taught me the most about how to act on climate change.

Steve Schneider was a brilliant climatologist, a kind human being, and a devoted global citizen. I had the privilege of working with him as his assistant, and as his graduate student. He was my father in many ways. We shared ethical convictions, intellectual passions, and a commitment to safeguarding our planet and future generations.

As I remember Steve, I want to thank him for introducing me to some of the strong, fierce, loving, tenacious, and giving aspects of myself. He helped me become more me, and he taught me what it means to believe in myself and to leverage myself for the good of all.

As part of my remembrance of Steve, sometimes I read through a dedication I wrote to him as part of the May 2011 issue of the Stanford Journal of Law, Science, and Policy.[460] Here's an excerpt:

"He was an extraordinarily talented communicator. He could condense complex climate science and policy issues into a five-minute speech using simple language and wise metaphors, and then proceed to inform, educate, and advise Congressional and Senate committees. I watched him speak to diverse audiences. About speaking, he always said, "Know thy audience, know thy self, and know thy stuff!" He always began with what the audience knew, and then guided them to what he knew, what was uncertain, and what he thought we should do. He was quick on his feet and enjoyed updating his presentations with new information. He loved communicating with people. He always included pictures of birds, other wildlife, and fun photos from his travels. He often had clever jokes handy.

Jokes aside, he was a fighter for truth. He courageously debunked pernicious skeptics who sought to spread misinformation about the status of climate science. In the last few years, Dr. Schneider told me he was spending about thirty percent of his time (and, I thought, losing precious sleep) to repeatedly set the story straight. Albert Einstein was correct when he said, "Great spirits have always encountered violent opposition from mediocre minds.""

You can read the full dedication online.[461] You can also read more about him, his books, presentations, and his research on his website.[462]

Steve taught me many lessons while he lived: have integrity, be optimistic, carry out and rely on good science, know your biases, be an ethical and proactive citizen, and above all, be a kind and loving human being. However, he also taught me something through his death. I learned that exhaustion could be fatal: Steve was a cancer survivor, but he was exhausted from being up late at night to debunk climate skeptics. I repeatedly asked him if it was worth it, since his lack of sleep was becoming harmful. His response was, "If I don't do it, who will?"

When Steve passed away unexpectedly, I was struck by the importance of slowing down. I realized that I, too, had been working incessantly and that it was taking a toll on my health. I began to give more time to meditation, being outside, and resting. It took me a few years, but I have changed myself to become more mindful and more balanced. My ability to cultivate balance through meditation has made me faster, more productive, and more creative.[463]

My transformation led to the development of Alchemus Prime,[464] which emphasizes authenticity and balance as we solve some of the planet's most daunting challenges by leveraging the best that science has to offer to be our best selves, and develop the most robust solutions possible.

I know that Steve knew the importance of his efforts, and that he wanted to give as much as he could in the time he had left. And he did. Everyone who knew him appreciates all he gave. In the end, he gave his very life. Steve's life and his death are priceless lessons for me and I carry them with me everyday.

I want the world to be the balance that I aim to embody with my everyday actions. A balance between doing and being, working and playing, giving and receiving, loving and letting go, empathy and detached compassion; a balance between self-care and selflessness. Because our most creative, effective, and brilliant ideas and solutions emerge when we take breaks, rest, and play.[465]

I know Steve is still on my team of mentors, encouraging me to keep going. It is through the healing of the inner self that we heal the outer world.[466]

Thank you Steve, for always being there. You continue to teach me so much. One of the priceless lessons has been to sustain myself in my battle against climate change.

Reflection:

How do you practice self-care and cultivate balance in your life? Why do you think self-care is important for climate warriors in particular?

Example from Marilyn: I take walks, meditate twice daily, draw, write poetry, and cook. I think climate warriors can become burnt out and dejected easily because of the doom and gloom in politics around climate change.

1.

2.

3.

Climate Change, Uncertainty, and Behavior

I often think about uncertainty in the context of climate change, and am a proponent of the precautionary principle.[467] Essentially, this is how it works: when you don't know the outcome, take actions that minimize risk as much as possible. In other words, not knowing the exact situation is no excuse for inaction, but a directive for preventive action. Applied to climate change, this would mean that we take all the necessary actions to minimize greenhouse gas emissions as quickly and effectively as we can.

Recently, I read an article[468] about uncertainty and human behavior, which yielded some interesting insights about the human condition and what author Jamie Holmes calls the need for closure. We want to be sure of a situation, a person, a decision, and so on. However, life keeps throwing uncertainties at us, and we must keep adapting. Holmes developed a quiz[469] that helps you figure out how tolerant you are of uncertainty – my result was "Master of Change," which is reassuring given I work with different clients in a constantly changing business environment.

With what we know about climate change, including the 2014 IPCC summary for policy makers (SPM) of the Fifth Assessment Report (AR5),[470] is that human-induced warming of the atmosphere is "unequivocal" and greenhouse gas emissions are the "highest in history." Given this situation, we must waste no time in implementing measures to reduce our emissions globally, and I have written extensively in Chapters 1 and 3 about what actions we must take. At the top of the list is switching to a plant-based diet, due to the harmful impacts of the meat and dairy industries.[471]

There is another approach that is high on the list as well. Canadian Prime Minister Justin Trudeau, while speaking at the annual meeting of the United Nations Framework Convention on Climate Change's (UNFCC), Conference of the Parties (COP 21) in Paris, touched on a very important strategy for addressing climate change: leveraging indigenous knowledge systems.[472] His argument, which I readily agree with, is that indigenous peoples understand how to: 1) live in harmony with nature, 2) adapt to changing conditions, and 3) use natural resources without destroying the planet.[473] These are all skills we desperately need right now to fight climate change, stabilize human population, and transition to a cleaner, safer, and more sustainable economy and lifestyle. Indigenous knowledge systems contain precious guidelines for diet, herbal medicine, architecture, agriculture, and many more systems. To understand and scale human resilience in the face of global climate change, then, requires that we leverage the best science on uncertainty and human behavior, and that includes the most appropriate, locally-attuned examples of adaptive behavioral responses from indigenous knowledge systems.

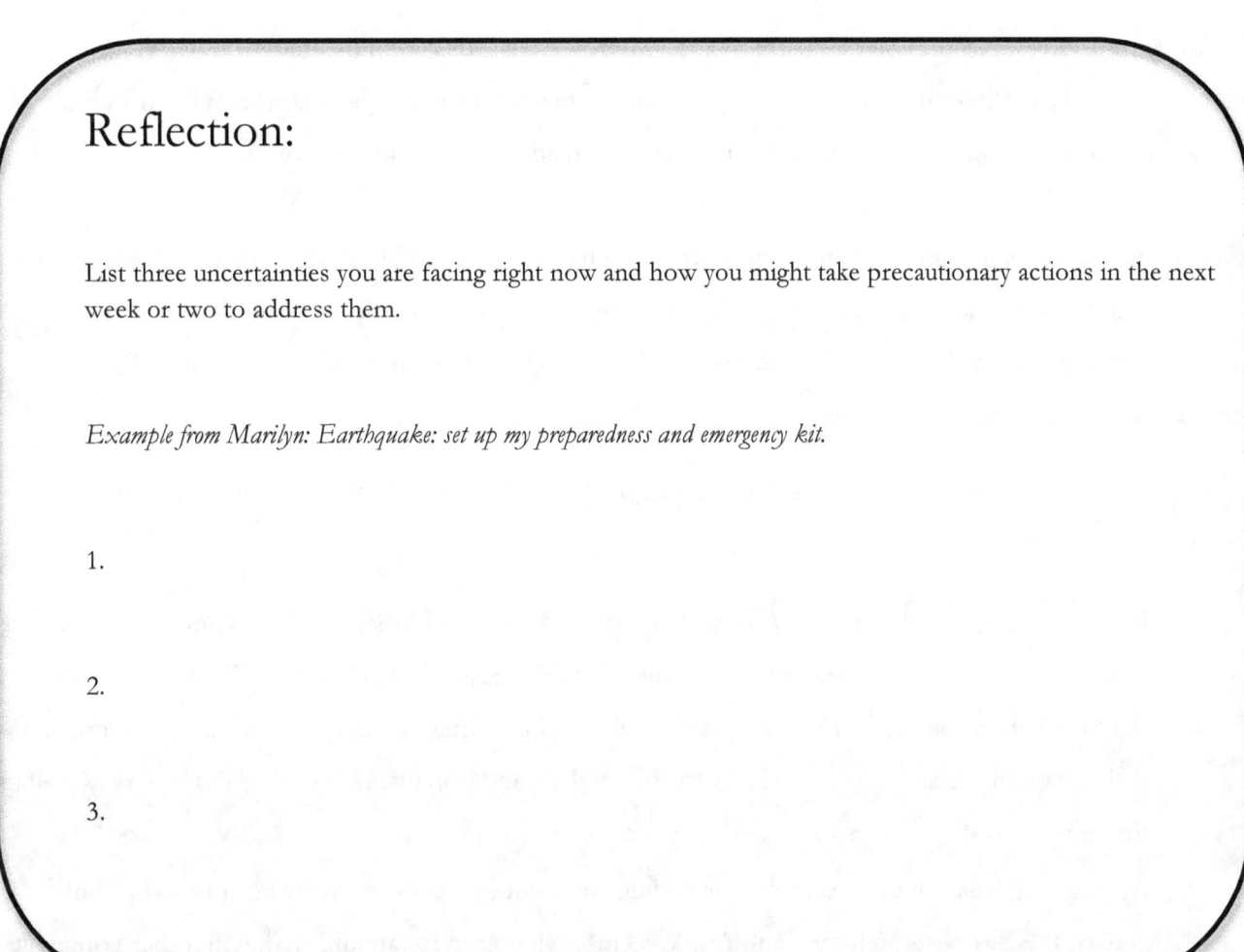

Reflection:

List three uncertainties you are facing right now and how you might take precautionary actions in the next week or two to address them.

Example from Marilyn: Earthquake: set up my preparedness and emergency kit.

1.

2.

3.

Companies Lead Clean Energy Efforts

As Apple transitions into the solar industry,[474] and Amazon signs a deal with its utility[475] to work more directly together to achieve its renewable energy goals, it's time to acknowledge that this is the beginning of a new era for energy: innovative corporate action toward mainstreaming clean energy.

For Alchemus Prime, this is a very affirming shift, as we have been coaching startups that will leverage and aggregate energy efficiency practices, renewable technologies, and rigorous behavioral sciences into a suite of services that will help clients achieve their clean energy goals, and much more in the realms of productivity, water, waste, and wellness.

Here's a summary of some of the juicier implications of this budding trend toward more corporate leadership in carbon-free energy:

- If large companies like Apple, Tesla, and SolarCity[476] begin to participate directly in the clean energy market, what does this mean for utilities? In the case of Amazon, its utility is still involved, but SolarCity is starting to behave a bit like a utility, indicating that corporations are experimenting with becoming clean energy providers for themselves and potentially for other customers as well in the future.

- Aggregating small energy loads, and providing clean energy access to many customers is a holy grail of the clean energy future, and Apple seems well poised to tap into that, with other companies sure to follow if Apple succeeds.

- When companies manage their own energy supply, they can link their rates to their energy usage, which means a more important and explicit role for behavior change in keeping costs down, since how much energy we use (a behavior) will have a direct impact on our bill (feedback).[477]

- As companies produce and consume their own clean energy, they have opportunities to reduce market risks and greenhouse gas emissions in their regions.

- This shift shows large companies are moving beyond green washing and lip service to actually working on achieving their renewable energy goals. The Amazon example is commendable, according to American Council On Renewable Energy (ACORE) President and CEO Gregory Wetstone, via GreenBiz:[478]

"Amazon, an active member of ACORE's U.S. Partnership for Renewable Energy Finance (US PREF), pledged in 2014 to go 100 percent renewable. Today it is making that pledge real by developing and integrating large amounts of renewable power onto the grid. This agreement gives flexibility to utilities to do their job in maintaining the grid, while also emphasizing the benefits of carbon-free energy and removing barriers to going 100 percent renewable. Amazon and Dominion are at the leading edge of clean energy development and together they are proving the economic and environmental value of corporate investment in renewables."

According to the New York Times, the possible merger of Tesla and SolarCity is another strong example of this trend:[479]

"'This is an effort to build the Apple of clean energy,'" said Daniel M. Kammen, the director of the Renewable and Appropriate Energy Laboratory at the University of California, Berkeley. "'That really is part of the new wave of companies that could make this decarbonization addressing climate change really work.'"

My late mentor and advisor, Steve Schneider[480] always said that in order to address the climate crisis, we must "do well by doing good." Steve was a strong proponent of partnerships that leverage innovation to help empower private and public sectors in the fight against climate change. As smart companies discover that doing what is good for the climate and planet is overall better for business and for people, they are beginning to drive this innovation. For Alchemus Prime, we are joyfully witnessing and participating in the fulfillment of our mission: demonstrating that it is good business to sustain nature.[481]

Reflection:

List three organizations you know of, or look up three organizations, that are leading the fight against climate change. How might you support one of them?

Example from Marilyn: The Center for Biological Diversity advocates for a dietary shift away from meat through their "Take Extinction off Your Plate program." I support them by publicizing their effort.

1.

2.

3.

Girl Scouts Rock Energy Savings

I remember back in summer 2010, contributing ideas to a brainstorm for specific actions girl scouts could take to reduce their energy use. We were beginning to design a study that would be a follow up to one of my collaborative dissertation studies: a curriculum for energy-saving behavior change in high schools.[482]

I'm excited to report that the GLEE (Girls Learning Environment and Energy)[483] study has been published[484] and is collecting accolades.

The study[485] is a rigorous experiment that investigates whether a social cognitive theory-based intervention (in this case, an energy conservation course with five modules) changes the energy-saving behaviors of girl scouts, whether the girls influence their parents to also change their behavior, and whether these changes persist over time (a median of almost 8 months).

Results[486] indicate that the intervention does change behavior, with 49% increased energy savings behaviors by the girl scouts, and through them, another 12% by their parents. This is very promising for future behavioral interventions, because the behavior change is directly caused by the intervention, and girl scouts form a large subset of the U.S. population with high penetration into households.[487]

If you're interested to look at the intervention materials, you can request them from lead author Hilary Boudet.[488]

Now, if only the girl scouts could sell healthier cookies[489] (my wish list: vegan, gluten-free, sugarless, and free of high-fructose corn syrup and hydrogenated oils). Healthy cookies will empower them to spread wellness across the country, and boost the food-related energy-saving behaviors in this and future studies. Then, they could address climate change and wellness challenges in tandem, as we aim to teach and practice at Alchemus Prime – yeah![490]

Reflection:

What are three community groups in your neighborhood that you could join to help reduce energy use and/or improve dietary habits?

Example from Marilyn: Neighborhood Watch, Block Party Committee, Carpool Club.

1.

2.

3.

Climate Leadership: Not a Moment to Lose

With the U.S. out of the Paris Agreement, several setbacks loom for global efforts to address climate change. According to climate scientist Michael Oppenheimer, these setbacks include:[491]

1. The imminence of a 2-degree C warming, which the Intergovernmental Panel on Climate Change, (IPCC) calls "dangerous climate change."[492]
2. Probable lack of transparency within the Paris agreement with the absence of the U.S., as China's tendency is to obfuscate; this would lead to lower accountability and inadequate results.
3. Hurting business by withdrawing uniform federal regulation, which would mean different state regulations that would complicate business operations.
4. Reducing the demand for clean technologies due to fewer incentives, although market prices continue to come down.

So far, the business sector has responded with concern and opposition.[493] Elon Musk, CEO of Tesla and SpaceX, withdrew his participation in U.S. presidential councils. Jeff Immelt, head of General Electric, has also expressed disappointment. A number of large companies, including Apple, Facebook, Google, Salesforce, and Microsoft, ran an advertisement stating their support for the Paris Agreement for sound business reasons.[494]

On a hopeful note, industry is already doing a lot to protect its bottom line, which it understands is connected to sustainability and climate change leadership.[495] Some notable examples:

*   Several Facebook data centers are powered by 100% clean and renewable energy.[496]
*   Apple's new campus runs mostly on solar panels.[497]
*   Walmart[498] and Exxon[499] are stepping up to reduce their climate impact.

Cities are also getting fired up, with the Climate Mayors group, comprising 88 mayors, writing to Trump[500] expressing its commitment to continue to lead by increasing investments in renewable energy. The United States Climate Alliance[501] has also formed, along with an unnamed group with several states, cities, and university presidents participating.[502]

As we face what is one of the worst presidential decisions ever, it is important to stay focused and fight even harder for climate stability. Businesses are doing just that.[503] There is not a moment to lose.

Reflection:

How will you assist your employer, city, university or alma mater, or state to fight climate change?

Example from Marilyn: Scale up my efforts through local, state, or international governments, through writing books and blogs, and by working more directly on wellness issues to reduce climate impact.

1.

2.

3.

Chapter 10 Notes

_____

[460] Learn more about the journal: https://journals.law.stanford.edu/stanford-journal-law-science-policy

[461] Read the full dedication here: http://web.stanford.edu/group/sjlsp/cgi-bin/orange_web/users_images/pdfs/61_Cornelius%20Dedication%20Final.pdf

[462] Steve's website: http://stephenschneider.stanford.edu/

[463] See Chapter 6 for the benefits of meditating.

[464] See Chapter 11 for how Alchemus Prime took form.

[465] Learn more about how to be in a flow state: http://www.fulfillmentdaily.com/workaholism-doesnt-work-embracing-flow-way-go/?

[466] This relates to the personal and planetary wellness framework that this book represents; see the Introduction and Chapter 1.

[467] Read more about the precautionary principle in this excerpt of the IPCC's climate mitigation report: http://www.ipcc.ch/ipccreports/tar/wg3/index.php?idp=437

[468] Read the full article here: https://www.theatlantic.com/health/archive/2015/10/the-benefits-of-getting-comfortable-with-uncertainty/409807/?utm_source=SFFB

[469] Take the quiz here: http://jamieholmesbooks.com/nonsense-quiz

[470] Read the summary for policy makers here: https://www.ipcc.ch/pdf/assessment-report/ar5/syr/AR5_SYR_FINAL_SPM.pdf

[471] See the essays in Chapter 3.

[472] Read more about this approach: http://aptnnews.ca/2015/11/30/trudeau-says-indigenous-people-can-teach-how-to-care-for-the-planet/

[473] Learn more about how researchers and indigenous people are working together to address climate change: http://news.unl.edu/newsrooms/unltoday/article/researchers-help-tribes-enhance-drought-and-climate-resilience/

[474] Read more about Apple's solar strategy: https://www.greenbiz.com/article/why-apples-new-energy-business-should-scare-utilities

[475] Learn more about how Amazon is partnering with its utility: https://www.greenbiz.com/article/amazon-and-utility-strike-breakthrough-renewables-deal

[476] See what Tesla and SolarCity are doing in the clean energy field: https://www.nytimes.com/2016/06/24/business/energy-environment/testing-the-clean-energy-logic-of-a-tesla-solarcity-merger.html

[477] See Chapter 9 for more on behavior change.

[478] See Note #411.

[479] See note #412.

[480] See the first essay in this Chapter for more on Steve Schneider.

[481] For more on Alchemus Prime's mission, see Chapter 11.

[482] See the study abstract: https://link.springer.com/article/10.1007%2Fs12053-013-9219-5

[483] Learn more about GLEE: https://glee.stanford.edu/

[484] Read the GLEE paper here:
https://www.nature.com/articles/nenergy201691.epdf?referrer_access_token=6lIGT24ycE7MLRUiP71Tet
RgN0jAjWel9jnR3ZoTv0OSI-td8XBozlbLMyUW4RtLK-
cPPq4D73BqlctaSDNSC3gr3sjpRQ_uyw0y5inWE2qgAnNV_TZt6I6861hJm8gE6nkzROOas38OKuOSW
VoZLPtQTZhdh-SNBLpfKJfjbfiG01asRCqgZUw0AX9eS1KyZU-
vQNPmJChgP7vCw_AbLBJGU2_N3fOmwqELHepFx5loVrIlP0Y4DNKF13Z6A3XtH3zVWF4jfpvhUP
QjW8ZkaA%3D%3D&tracking_referrer=www.latimes.com

[485] Ibid.

[486] See the L.A. Times story about the GLEE study: http://www.latimes.com/science/sciencenow/la-sci-
sn-girl-scouts-energy-conservation-20160711-snap-story.html

[487] Read more about the girl scouts: http://www.girlscouts.org/en/about-girl-scouts/who-we-are.html

[488] Request the full study materials here: https://healthimprovement.stanford.edu/GLEE/

[489] See this article about girl scout cookies: http://www.cnn.com/2015/04/21/opinions/maizes-girl-scout-
cookies/

[490] See our website for how we work to address climate change and wellness challenges together through
behavior change: http://www.alchemusprime.com/how-we-work/

[491] Read the full interview with Michael Oppenheimer:
https://www.theatlantic.com/science/archive/2017/06/oppenheimer-interview/529083/

[492] Read more about dangerous climate change:
https://www.ipcc.ch/publications_and_data/ar4/wg3/en/ch1s1-2-2.html

[493] Read about the business community's reaction to Trump's withdrawal from the Paris agreement:
https://www.usnews.com/news/national-news/articles/2017-06-01/businesses-balk-as-trump-pulls-out-of-
paris-climate-agreement

[494] See the ad here: https://www.c2es.org/international/business-support-paris-agreement

[495] See what the business community is doing to fight climate change:
https://www.wired.com/2017/06/even-without-paris-business-will-leave-trump-behind-climate-change/

[496] Read more about Facebook's data centers operating on clean energy: https://sustainability.fb.com/clean-and-renewable-energy/

[497] Learn more about Apple's new Silicon Valley campus: https://www.wired.com/2017/05/apple-park-new-silicon-valley-campus/

[498] See what measures Walmart is taking: http://news.walmart.com/2017/04/19/walmart-launches-project-gigaton-to-reduce-emissions-in-companys-supply-chain

[499] See what Exxon is doing now, with a change in leadership: https://www.reuters.com/article/us-exxonmobil-climate-idUSKBN18R0DC

[500] Read their statement here: http://www.sandiegouniontribune.com/opinion/the-conversation/sd-cities-states-against-trump-climate-decision-20170601-htmlstory.html

[501] Read more about this alliance: http://ens-newswire.com/2017/06/04/states-cities-intensify-climate-protection-efforts/

[502] Learn about this new group: https://www.nytimes.com/2017/06/01/climate/american-cities-climate-standards.html?_r=0

[503] Read even more about what businesses are doing to fight climate change: http://www.economist.com/news/business/21723160-american-firms-can-offset-donald-trumps-pullout-paris-big-business-sees-promise?fsrc=scn/fb/te/bl/ed/bigbusinessseesthepromiseofcleanenergy

Chapter 11: Leading With Integrity

Identity: What is Your Point of Reference?

While meditating one morning, I remembered some research[504] I had done as a grad student on identity. Specifically, 'global identity,'[505] which refers to:

"consciousness of an international society or global community transcending national boundaries, without necessarily negating the importance of state, nation, or domestic society"

Originating in United Nations (UN) literature, global identity has been highlighted to motivate action against global conflict, poverty, epidemic diseases, and environmental degradation. I applied the concept to energy use and climate change, and found that identifying globally is associated positively with our energy use reduction behaviors.

Of course, we have many identities; one or more are salient at any given time. Do you identify as a human being, vegan, mother, father, brother, sister, friend, activist, academic, inventor, professional, artist? The list goes on and on.

In my meditation, the link between my own global identity and my point of reference became apparent. Since high school, I've had a very strong sense of global identity; I suspect it came from reading about the world's ills and feeling a passionate commitment to do something about them. When I thought about what I wanted to do in the world, I thought about what the world needed and how I could make a sizable impact. This way of thinking led me to study environmental resource management, and then climate change and behavioral change. I also studied and taught design thinking, attracted by its intrinsically motivating process. I worked for the UN on multiple environmental projects in ten Pacific Island Countries, and then for myself, co-creating and applying a model for changing the way teams, organizations, and communities collaborate and solve problems.

Another tendency I have is to put others first, especially family, friends, and the disadvantaged living beings of the world. I identify strongly as a helper and giver. I lost sight of myself here too, prioritizing their needs above my own. After many years of working in this way, I succumbed to adrenal fatigue. With facilitation and guidance from my mentors, I experienced a deep healing and was able to see myself from a new (and more accurate) perspective.

I began to use a new point of reference: me.

Somewhere along the way, in all my thinking about the Earth, about animals who suffer, and about other people, I had lost sight of myself. The insight that emerged once again for me to absorb and never forget was this: If I put myself first and meet my own needs, I will be better able to serve the world, give to the world, and benefit those who need my help.

A close friend said something to me almost a decade ago that now comes back to me, and puts this insight in another way:

"Do what makes you happy, and the universe will be happy."

Thanks, Chien-Wen, you are correct!

This insight transformed my life; it was time to learn the lesson once and for all. Of course, I'm still learning it and it will likely take a lifetime, but at that point, I stopped what I was doing in order to stop the way in which I was doing it, and changed my point of reference to me. I could feel my neurons rewiring and re-firing. Now, I begin with my needs, my sensitivities, my talents, my goals, and my unique perspectives. I begin with grounding, meditation, dream journaling, and reflection; connecting with Reiki energies and the source of my intuition, I am guided at every step. I cultivate clarity about what I want, and how I want to work, give, and serve. It's time to begin again with an approach that honors me, so that I can serve from the best of me. Every so often, I become fixated on helping others, but return to my own sustenance, which helps me serve others even better.

Intuitively, I realize that I now listen to my own voice; all my love for the world is being heard louder and clearer than ever before. I am grateful to rediscover myself in my quest for global transformation. The change I initiate in me, I bring to the world.

What are the identities that are salient for you? What is your point of reference? Are you putting yourself first? Do you serve your causes from the best of you?

Reflection:

Who and what do you typically put before yourself when you prioritize? Why? How might you put yourself first?

Example from Margaret: I put my children first, due to the deep sense of responsibility I feel for bringing them into this world. Now that they have grown up and are independent, I can put myself first. For example, I engage in my hobbies, including traveling, sewing, and other creative projects.

In Search of the True Self: Alchemus Prime

In 2014, meditation proved to be the most important activity in my life. Soon enough, I noticed many insights and inspirations arriving during my daily meditations.

Towards the end of the year, life gave me a major transition: closing down my previous company, d.cipher, and embarking on a new beginning. Aside from meditation, there was one other force on my side: serendipity. Exhibit A: I met my mentor a few months before I realized that my previous company would need to be shut down, and he helped me make a new start. Exhibit B: I met a stranger on a plane I wasn't supposed to be on, at 1am, who led me to the inspiration that has ultimately manifested as my new company name.

It all began, however, in the Navajo Nation in December, 2013. During a full moon, around midnight, at a sacred ceremony, the men's chanting took me to a place where there was nothing else: only inspiration.

Since then, the pieces have begun to fall into place. What was missing before has arrived and seamlessly integrated itself into the conceptual model for my new company. I started my consulting career facilitating change management and creativity workshops for leaders in climate change, wellness, and education. This was most rewarding, except there was a very important aspect of leadership, perhaps the most important aspect, that wasn't being included explicitly: integrity.

The epiphany arrived gently during one of my meditations late last year: if meditation was so important, *why* wasn't it part of what I was offering to my clients?

Meditation itself was missing from my approach! How else might we have the greatest positive impact if we aren't in tune with our real, true, and full selves?

I began to formulate a new model, taking what I already had:
- behavioral sciences (so we can effectively change our behavior and practices to meet our goals),
- design thinking (so we can consistently be innovative in the way we solve problems),
- biomimicry (so we can emulate nature's elegance and balance to create sustainable solutions),

and melding it with meditation, the art of being in the present moment, in touch with oneself, and centered in integrity and authenticity.

Ponder this: we know from quantum physics that the observer changes the observed. So if we are going to make an impact on the world, why not begin from our true selves so that we can make the most positive and genuine impact?

The model took shape and I started calling it the diamond model – inspired by the four directions and the four sacred mountains in the Navajo tradition, and the fact that the diamond is the strongest material substance on earth. At first, I called the four components Intuition, Integration, Innovation, and Transformation (which also led me to start blogging). Later, these names evolved into connect (meditation), create (design thinking), transform (behavioral sciences) and sustain (biomimicry).

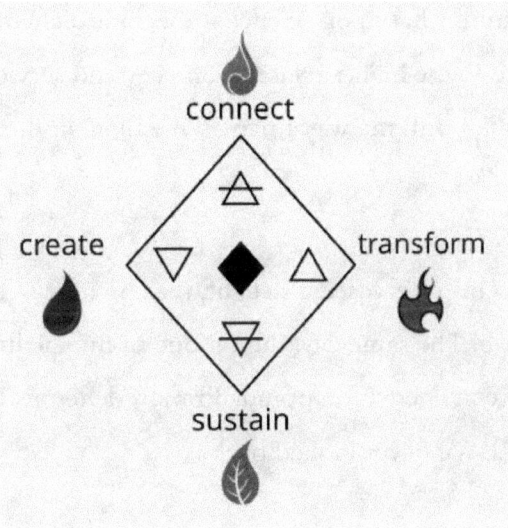

The Alchemus Prime Diamond Model leverages the strengths of four bodies of knowledge into one integrated whole.

Then, in January, I traveled to Ecuador. On the way, having missed two flights and having barely made it onto my third plane, I met a stranger who inspired me immensely. He consults with clients to help them balance their masculine and feminine energies, something I believe is crucial for the future we are co-creating. At 30,000 feet, traveling toward a new sunrise, he and I bonded over many topics; one of them was books. Later, we wrote lists of the books that changed our lives. One of the books on his list caught my eye: *The Alchemist*, by Paul Coelho.

I was struck by the fact that I hadn't yet read this incredibly famous book. I immediately purchased it and read it in one sitting. All-over-my-body-goose bumps came, and stayed…

Soon after this experience, it was time to go on retreat with my mentor to birth my new venture. I was so inspired by *The Alchemist* that I chose the desert as my retreat venue; we went to Joshua Tree National Park.

There, after a couple of days of intense brainstorming and visioning, I went for my first sound bath at the Integratron[506] and felt a spaciousness in my mind. Right after that, we went to the national park, where I wandered, communing with the teeming life in the vast desert. Inspired by nature, we returned to our work, now generating ideas for company names. After our chart paper was full of more than two hundred ideas, my mentor suddenly said: "Alchemist Prime."

I started to smile and realized that this was poetic justice: this name encapsulated what I wanted to give my future clients: a structured opportunity based on science to become alchemists of their own character and actions for a better planet. To rise to the highest state of efficacy and service as leaders, and make a difference. And, the name resonated with the way I named my blog, in the vein of my hero, and leader of the Autobots, Optimus Prime.

A week later we realized, much to my effervescent delight, that Alchemist Prime[507] *is an actual Transformer,* who looks to nature for inspiration! The same day, during one of my meditations, a new company name came to me, which was even more in line with Optimus Prime: Alchemus Prime. My model became the Alchemus Prime Diamond Model. We never looked back.

We also observed the seamless synergy between our approaches, and began a collaboration that proves to be delightfully creative, fearless, and evolving. We are in this for the win-win solutions that benefit our clients and the world. Alchemus Prime is about us all, doing alchemy together, to reach our prime potential as leaders called upon to protect and nourish a world in chaos and transition. We stand on the foundation of science, combining behavioral tools, processes for creativity, biomimicry principles, and the ancient and scientific techniques of meditation to hone a whole person, a well-rounded leader, a steward of the future. That's you, and me. That's us.

For our logo, we chose the yin yang, symbolizing the interconnected nature of opposing forces, including the masculine and feminine energies, and replaced the circles with a diamond shape to symbolize the

Alchemus Prime Diamond Model. We combined this image with the four elements, air, water, fire, and earth, as typically represented in alchemy.

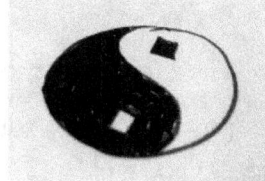  

Iteratively designing the Alchemus Prime logo.

Our vision is to align business with nature so that all life can flourish. We are facilitating experiences that will activate the true self to empower our clients to change the world in beautiful ways. As I continue on this exciting path, I am still on my way to finding, being, and honoring my true self in service of personal and planetary wellness. Won't you join me?

Reflection:

Describe your true self.

Example from Marilyn: My true self is fiercely devoted to people and planet. I am creative, analytical, and integrative. I love animals, flowers, trees, and the ocean. I am giving, loving, and nurturing. I love to cook, walk, write, and facilitate. I am whimsical, goofy, and nerdy. I love the Transformers and Star Wars. I always aim to do the right thing.

Science: Vulnerability is Beautiful

According to social scientist Professor Brené Brown,[508] humans are wired for connection. "It's why we're here," she says in her TED talk.[509] In fact, the need to connect is so powerful that any incident of disconnection brings shame to us, and a sense of "excruciating vulnerability." Further, her research shows that those who have a strong sense of love and belonging, *believe* they are worthy of that love and belonging. This belief is the single most important variable separating those with high levels of love and belonging and those with low levels.

Prof. Brown calls people with a high sense of love and belonging "whole-hearted" people, and their key characteristics are courage, compassion, connection, and *vulnerability*. For these people, vulnerability is not excruciating. Rather, vulnerability is beautiful. Prof. Brown goes on to say that vulnerability is the "birthplace" of joy and creativity.

According to Prof. Brown's research, we cannot selectively numb feelings – when we try to numb shame or sorrow for instance, we also numb joy. The more afraid we are, the more we cling to certainty and perfection when there isn't any. How might we navigate our emotions and vulnerability to face reality and thrive? A must-watch talk! And, you can look up information on Prof. Brown's peer-reviewed publications[510] and books[511] as well.

What does this have to do with you? Well, being your authentic self, being in integrity, and leading from that place of the true self,[512] requires these same qualities: courage, compassion, connection, and vulnerability. Adapting to change, as I discussed in Chapter 7, requires connecting with others. Vulnerability in leadership requires that we conquer our fears and face our challenges in humble, sustainable, and exemplary ways.

Reflection:

How have you practiced or how might you practice being vulnerable as a leader? What has happened or might happen as a result?

Example from Marilyn: I show my whimsical and nerdy side when I lead workshops. I've done ridiculously silly things like pretend to be Optimus Prime or Yoda during ice-breaker exercises and introductions. I find this approach gives participants permission to relax and play, and their solutions are much more creative and effective as a result of this relaxation and fun.

Reflections: Good Leadership

For a while, I've been somewhat peeved about drought management in my beloved home state of California. We've been in a drought for at least four years and you'd think we would have gotten our act together by now. This situation prompted me to ask: what does good leadership look like?

If you recall, Governor Brown asked for reductions in household water use. Given that household water use accounts for 4%[513] of the state's water use, it's important to look at the other 96%. A whopping 80% of our water is used by agriculture,[514] and yet farmers aren't being asked to cut back.[515, 516] One of the top uses of water in agriculture is to grow alfalfa[517] for local and Asian meat and dairy production. Logic (and good, progressive, strategic leadership) would suggest, as the New York Times does, that we Californians examine our diet, and consciously shift away from meat and dairy[518] in favor of less water-intensive plant-based diets in order to free up water for essential needs such as drinking.

On one occasion I was at Stanford University for a meeting. I noticed the sprinklers coming on and was unpleasantly surprised. A close friend informed me that the water is sourced from a lake, and is not grey water.[519] I found this further surprising, coming from such an innovative institution, clearly perceived as a sustainability leader in Silicon Valley,[520] and my alma mater. Good leadership, in my humble opinion, would entail planting beautiful drought-resistant plants all over campus. Leadership isn't always conventional manicured prestige; leadership is pioneering a new look that exemplifies pragmatism, simplicity, humility, and adaptation.

When I participated in the inaugural Biomimicry for Social Innovation Immersion Workshop,[521] my esteemed teacher and colleague Toby Herzlich indicated that good leadership also means stepping back from what we usually do and taking a different perspective. For example, I'm one of those people who speaks up unabashedly in classes and workshops. For me, then, it is an act of leadership to step back, listen, and allow others who typically don't speak to practice speaking up.

So, good leadership is about being humble,[522] effective, and scientific; it's about observing, adapting,[523] being mindful[524] and curious; it's about examining and taking on different perspectives;[525] and it's about having a beginner's mind.[526]

With Toby and twenty-five other colleagues, I learned that good leadership, with nature as model, mentor, and measure, also includes integrity, decentralized authority, minimal or no hierarchy, cooperative relationships, simple rules, and tight feedback loops – these are covered in Chapter 7. Above all, good leadership means creating solutions that uplift and sustain life, the way nature does.

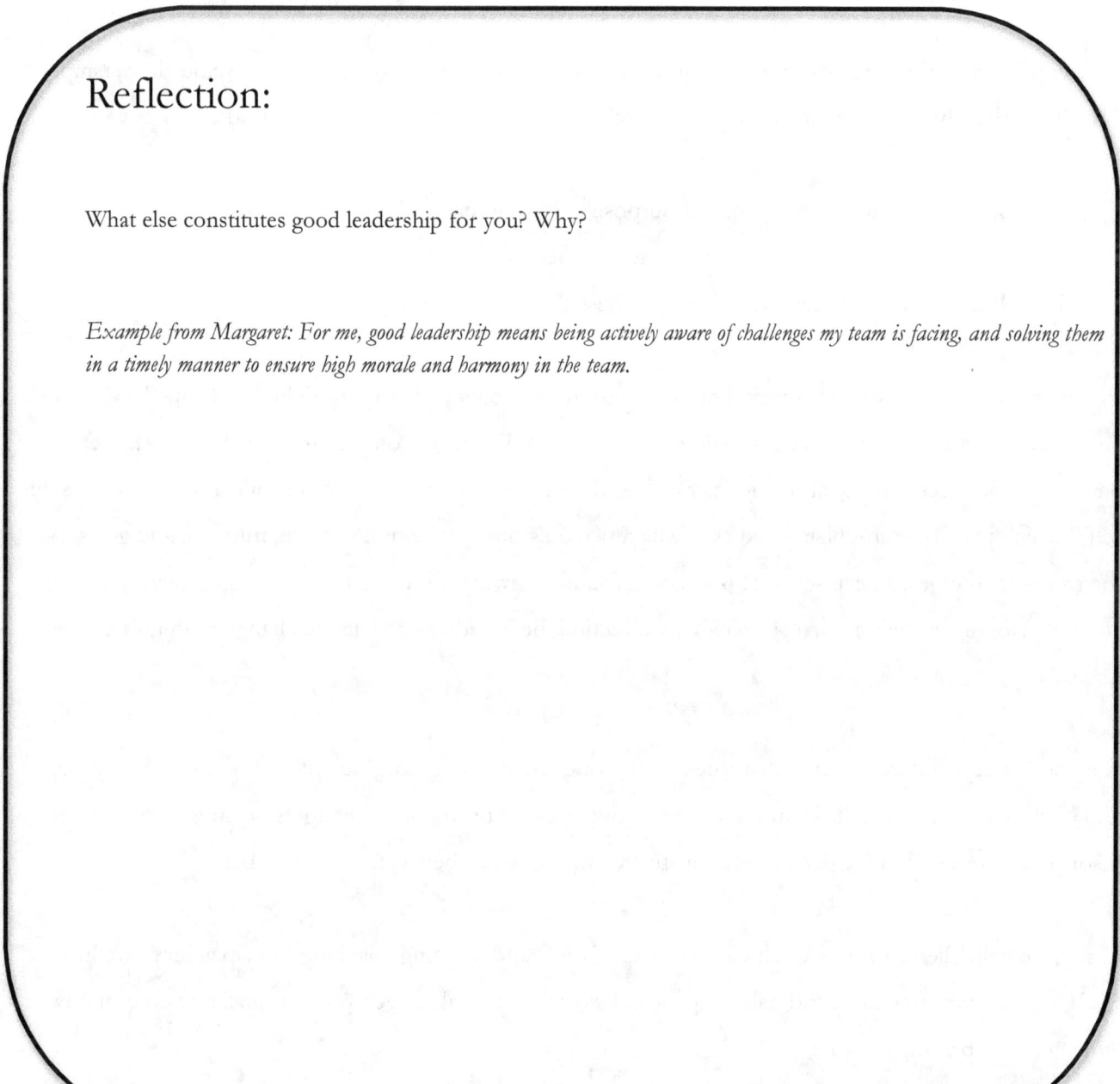

Reflection:

What else constitutes good leadership for you? Why?

Example from Margaret: For me, good leadership means being actively aware of challenges my team is facing, and solving them in a timely manner to ensure high morale and harmony in the team.

Conscious Leadership = Happiness

I've been following Emma Seppälä,[527] Science Director of the Center for Compassion and Altruism Research and Education at Stanford University, for over two years. I really like her approach to the integration of wellness and productivity. I've read her excellent book, *The Happiness Track*[528] – check it out!

Dr. Seppälä has written an article that is particularly important, about what I'll call conscious leadership.[529] She claims that three factors, according to research, are critical for "extraordinary" leadership:

1. Inspiration: embodying a sense of purpose and meaning.
2. Kindness: being thoughtful and sincere to others.
3. Self-care: systematically prioritizing wellness despite being busy.

In my own experience with clients and myself, these traits are indeed crucial. Alchemus Prime has facilitated about 20 executive retreats, and clients have thanked us for our focus on the alignment between personal and professional goals using science-based tools. In our approach, we include a focus on "the true self," which involves scrutinizing and clarifying a person's or organization's vision, mission, and goals. As the true self emerges, a deep sense of purpose is found, allowing our clients to orient themselves mentally and behaviorally in their chosen sustainability direction, be it addressing climate change, water, or energy issues.[530]

We also foster a culture of being that true self at work, and creating non-hierarchical bonds between staff and management that facilitate smoother operations because protocols and targets begin to carry deeper resonance, grounded in the personal connections and alignment between team members.

Lastly, we help clients prioritize self-care, providing tools and ongoing coaching to help achieve wellness goals, create more free time, and balance personal, social, financial, career, professional, and community or "giving back" priorities.

Reflection:

How do you experience or cultivate non-hierarchical ways of connecting with people around you?

Example from Marilyn: I say thank you a lot, give credit where it's due, and ask for advice from those who report to me. In workshops, I use playfulness and cross-functional teams to dissolve hierarchy.

McKinsey: Four Key Leadership Behaviors

At Alchemus Prime, we take a behavioral approach[531] to, well, everything. So, encountering recent research from McKinsey that talks about four critical leadership behaviors was music to my ears!

McKinsey and Company surveyed[532] 189,000 people in 81 organizations around the world, and found:

"…leaders in organizations with high-quality leadership teams typically displayed 4 of the 20 possible types of behavior; these 4, indeed, explained 89 percent of the variance between strong and weak organizations in terms of leadership effectiveness…"

Here's an analysis of McKinsey's top four leadership behaviors from the perspective of the science-based Alchemus Prime Diamond Model:[533]

1. **Effective problem-solving:** being able to solve a problem hinges on being able to accurately recognize, define, and frame the problem. We use design thinking[534] to reframe challenges to ascertain what the root causes are, and whose perspectives matter, and then use integrated behavioral, design thinking, biomimicry, and meditation tools to help resolve the core issues. We take this approach because inevitably most problems our clients encounter have some mix of interacting social, economic, environmental, and personal aspects.

2. **Strong results-orientation:** at the end of the day, process is crucially important *in order to deliver results*. We leverage design thinking to maintain a high level of mindfulness of process, and help clients with proven behavioral tools to keep a clear focus on results based on SMART goals[535] and quantified metrics that stem from strong core values, and a vision and mission that reflect the team's authentic identity. We go to the root of the leadership team's integrity and work from there to ensure ethical, structural, and procedural alignment in the organization, which forms the foundation of high quality results.

3. **Different perspectives:** Strong leaders welcome and consider multiple perspectives. We structure this into processes with our clients using biomimicry, mindfulness, behavioral, and design thinking tools. With our guidance, and using our 360-degree win-win perspective, insights gleaned from varied perspectives allow leaders to make decisions that best serve their clients, stakeholders, employees, and the Earth.

4. **Support for team members:** We nurture strong, trusting, and resilient relationships that endure. Trust is a key ingredient for creative and cohesive teams. We help team members connect authentically using guided mindfulness exercises; nature immersion; and creative, analytical, and team building processes. We choose clients carefully based on their core values, passion for service to people and planet, and integrity, so that we can build trust.

It's affirming to find our approach in alignment with this research on leadership, as we are a leadership company focused on addressing climate change and wellness problems together through behavior change integrated with other science-based tools.

Reflection:

How do you define good results personally and professionally?

Example from Marilyn: Personally, the result is good if it's in line with my core values, and if I'm happy in the process and with the outcome. Professionally, a result is good if it solves the challenge or problem at hand without creating any other challenges, and even better if it provides side benefits.

Chapter 11 Notes

---

[504] Read more about my research on global identity: http://web.stanford.edu/group/peec/cgi-bin/docs/behavior/research/global%20identity%20manuscript%20final.pdf

[505] Ibid.

[506] Learn more about this unique structure: https://integratron.com/

[507] Learn more about Alchemist Prime: http://tfwiki.net/wiki/Alchemist_Prime_%28Prime%29

[508] Read more about Professor Brown: http://brenebrown.com/about/

[509] Watch the TED talk: https://www.ted.com/talks/brene_brown_on_vulnerability

[510] Professor Brown's publications: http://www.uh.edu/socialwork/_docs/faculty-CVs/Brown.pdf

[511] Professor Brown's books: http://brenebrown.com/books/

[512] See other essays in this chapter.

[513] Learn how California's water is being used: http://www.onegreenplanet.org/news/californias-drought-whos-really-using-all-the-water/

[514] Learn more about agricultural uses of water: http://pacinst.org/publication/ca-water-supply-solutions/#issuebriefs

[515] Farmers are not being asked to cut back on water use: https://www.washingtonpost.com/blogs/govbeat/wp/2015/04/03/agriculture-is-80-percent-of-water-use-in-california-why-arent-farmers-being-forced-to-cut-back/

[516] I wrote this essay in 2015. A more recent look at the drought situation shows that, despite imprecise measurement, the Sierra Nevada snowpack is much larger as of spring 2017 compared to 2015, ensuring a better water supply situation: https://www.nytimes.com/interactive/2017/03/22/us/california-measuring-snowpack.html

[517] Learn about alfalfa production and water use: http://www.slate.com/articles/technology/future_tense/2014/05/_10_percent_of_california_s_water_goes_to_almond_farming.html

[518] Read the New York Times article here: https://www.nytimes.com/interactive/2015/05/21/us/your-contribution-to-the-california-drought.html

[519] Learn more about this water source: https://suwater.stanford.edu/water-supply-overview

[520] Learn how Stanford is leading in the sustainability arena: https://www.theguardian.com/sustainable-business/standford-university-students-change-world

[521] Learn more about this workshop here: https://biomimicry.net/what-we-do/professional-training/immersion-workshops/social-innovation-workshop/ and in Chapter 7.

[522] Read more about humble leaders in this article from Harvard Business Review: https://hbr.org/2014/05/the-best-leaders-are-humble-leaders

[523] Read more about adaptive leadership: https://www.forbes.com/forbes/welcome/?toURL=https://www.forbes.com/sites/travisbradberry/2012/11/09/leadership-2-0-are-you-an-adaptive-leader/&refURL=&referrer=

[524] Learn about mindful leadership: http://www.billgeorge.org/articles/mindful-leadership-compassion-contemplation-and-meditation-develop-effective-leaders/

[525] Learn about perspective taking: http://www.reachingresults.com/the-science-of-perspective-taking

[526] A beginner's mind is a powerful leadership tool: https://www.forbes.com/forbes/welcome/?toURL=https://www.forbes.com/sites/mikeotoole/2014/09/16/the-beginners-mind-how-naivete-can-become-a-critical-business-asset/&refURL=&referrer=

[527] Learn more about Dr. Seppälä: https://www.psychologytoday.com/experts/emma-m-sepp-l-phd

[528] Read *The Happiness Track*: http://www.emmaseppala.com/book/

[529] Read the article here: https://www.psychologytoday.com/blog/feeling-it/201601/3-things-extraordinary-leaders-do

[530] Read more about the true self in the rest of this Chapter.

[531] See our website for how we apply behavioral sciences: http://www.alchemusprime.com/behavioral-sciences/

[532] Learn more about the survey: http://www.mckinsey.com/global-themes/leadership/decoding-leadership-what-really-matters

[533] Learn more about the model here: http://www.alchemusprime.com/model/

[534] See Chapter 8 for more about design thinking.

[535] SMART goals are constrained in ways that make them easier to track and achieve: http://hrweb.mit.edu/performance-development/goal-setting-developmental-planning/smart-goals

Glossary

Alchemy – a change of state or transmutation e.g. from metal to gold

Alchemist - person who practices alchemy

Aquifer – a body of water underground that can provide a supply of drinking water

Behavioral sciences – a set of social sciences that help us understand and change behavior, such as psychology, social psychology, sociology, anthropology, and behavioral economics

Biodiversity - the variety of wildlife in nature, usually measured by the numbers and types of species in any given habitat or ecosystem

Biomimicry – emulating nature; a discipline that teaches us how to learn from nature

Bioneers – biological engineers; also the name of a conference focused on biomimicry

Capstone class – a multifaceted course that typically occurs at the end of an educational program, tying together and testing intellectual and other types of learning experiences such as leadership, team-building, interdisciplinary problem-solving, etc.

Cognitive dissonance – when our thoughts and actions are inconsistent and contradictory to each other

Decarbonization - reducing or removing the carbon content from something, e.g. energy production

Design principles – guidelines we can use to design a structure, solution, or process

Design thinking – a systematic process for creativity, taught by the company IDEO, and many design schools including the d.school at Stanford

Dichotomy – showing two things as having high contrast or difference, e.g. science and art

Epiphany – a sudden realization about the deep meaning of something, a sudden insight

Greenhouse effect – heating of the earth's atmosphere due to greenhouse gases becoming trapped in the atmosphere

Greenhouse gases – gases that absorb the sun's radiation and trap heat in the earth's lower atmosphere

Integrated methodologies – combining methods e.g. behavior science methods and design thinking methods, for a more holistic approach to problem solving

Integrity – aligning values, thoughts, speech, and actions

Iteration – to carry out a task or process over and over, usually to reach a stronger conclusion each time

Life's Principles - a set of principles for how to learn from nature, using the discipline of biomimicry, as taught by organizations such as the Biomimicry Institute and Biomimicry 3.8

Meditation – focusing the mind using a tool e.g. a mantra, visualization, or observation

Mindfulness – focusing one's attention to the present moment without judgment

Mutate – genetically change or adapt in structure or characteristics, usually in evolution

Obfuscate – to confuse, bewilder, or make something less clear

Serendipity – experiencing positive outcomes by coincidence

Social biomimicry – a term coined by Marilyn to describe the nature-conserving actions of indigenous peoples, and ways in which human biology is similar to ecology, suggesting deep connections between human nature and the rest of nature

Solar photovoltaic – technology that converts sunlight directly into electricity

Veganism – refraining from eating and otherwise using any animals or animal products

Further Reading

If you liked this book, try one of our compilations of recipes or poetry. More information is available on our website.[536]

1. *Food of Love: 39 Recipes from my Heart to Yours.* A plant-based cookbook containing 39 of Marilyn's favorite recipes, with personal stories and photos.

2. *One Friend:* Poems written from the perspectives of animals affected by climate change, animals in factory farms, and animal rights activists.

3. *Earth Champions: Birthing a New World.* Poems about the true self, overcoming obstacles, being one with nature, and cultivating love for all.

4. *Visionaries: My Life-Changing Mentors.* A set of poems about two of Marilyn's mentors and their profound influence on her life and work.

5. *Beyond Blood: Soul Connections.* Poems about connections we forge with other human beings, including family, friends, and beyond.

6. *Seasons of Life: Nature's Nurturing Energy.* Various moments of sacredness in nature, captured through poetry.

7. *Meditation: Inspired Moments.* A collection of poems about various approaches to meditation, including mindfulness, Reiki, and meditating in nature.

8. *Personal and Planetary Wellness: Addressing Climate Change, Wellness, and Social Justice Challenges.* A collection of essays about how to achieve wellness and address climate change using science-based tools. (The one you're reading).

9. Forthcoming: our second cookbook, *World of Love*. A plant-based and gluten-free cookbook with recipes from around the world. Stay tuned!

---

536 See: http://www.alchemusprime.com/our-books/

ABOUT THE AUTHOR

Marilyn is an engagement specialist, facilitator, and coach who works with leaders, teams, organizations, and communities to help address climate change and wellness challenges. Through her company, Alchemus Prime, she has implemented workshops, retreats, projects, and research that empowers integrated solutions. Marilyn learns from the behavioral sciences, design thinking, biomimicry and meditation techniques, integrating them in innovative ways using the Alchemus Prime Diamond Model. Aside from this collection of essays, Marilyn has authored six collections of poetry, and is currently working on her second cookbook and two novels. All of Marilyn's work centers on the theme of how to leverage individual and community action to address global challenges by making visible the deep links between personal and planetary wellness.

www.ingramcontent.com/pod-product-compliance
Lightning Source LLC
Chambersburg PA
CBHW081111170526
45165CB00008B/2416